Brooklyn, Buck Rogers
and Me

Brooklyn, Buck Rogers and Me

ERIC JOHN BISHOP

Writers Club Press
New York Lincoln Shanghai

Brooklyn, Buck Rogers and Me

Writers Club Press
an imprint of iUniverse, Inc.

For information address:
iUniverse, Inc.
2021 Pine Lake Road, Suite 100
Lincoln, NE 68512
www.iuniverse.com

ISBN: 0-595-26816-1

Printed in the United States of America

To Wanda,
who bestows infinite love on me,

and my father,
who believed his sons could do anything,

plus my colleagues,
who have shared these sometimes rocky paths.

ILLUSTRATIONS

Space and underwater photographs are available through the courtesy of General Electric Company Inc.; Fairfield, Connecticut.

Computer product photographs are available through the courtesy of Seagate Technology, LLC; Scotts Valley, California.

FOREWORD

Someone more important than I should write this introduction to Eric's book—a Flight Director at NASA, for example, or maybe an archangel. Eric would have liked that.

For years, he sat in the green chair in my living room on Tuesdays, drinking decaffeinated coffee and working through the making of this book. It was his life. Around him in the room were seven other writers who loved him, and who held his hand while I shouted at him to shorten his sentences and tighten the prose. He wrote and rewrote, rearranged paragraphs a hundred times, and dedicated the finished product to his wife, Wanda, to his father who is a part of the book, and to us.

So here it is, *Brooklyn, Buck Rogers and Me.* In earlier stages it sported other titles—but it was always the story of a restless New York kid who spent his life searching for what felt right. The last time I saw him, four days before his soul spun off to another plane, he said, "You know, I've read it's possible that people never die—they just live in parallel universes. We're actually all here at the same time."

"Ah," I said, over my chicken sandwich. "No wonder the world feels crowded."

He laughed and paid for my lunch, and told stories about his growing up years, as he was wont to do at these gatherings. He told us about his children, and his grandchildren whom he adored, about his experiences as an engineer, as a father, as a writer, and as a man.

This book is a compilation of those things. It's about friends he collected and projects he worked on, his beloved Brooklyn, his parents, and the Great Depression. His memories and photographs are priceless, so, at this moment, you hold a treasure in your hands. These pages recount his contributions to space exploration; you'll witness the author's love for the shape of the future.

He wrote, "It never hurts to have the pure dumb luck to be in the right spot at the right time. Be careful what you wish for though. You might get it."

What a stunning benediction.

Carolyn D. Wall

PROLOGUE

"The optimist proclaims that we live in the best of all possible worlds. The pessimist fears that this is true."

—James Branch Cabell, 1879–1958

For a man with a ninth grade education, my father had a unique appreciation of history and world events. Once I asked him, "You were seven years old when the Wright brothers flew, and at seventy-three you saw astronauts walk on the moon. What are your thoughts about that?"

Dad gave me an enigmatic smile and replied, "You ain't seen nothin' yet. Changes will come even faster in your lifetime."

His statement returns to mind with every new scientific achievement I see. Three decades after Dad's comment, there is more computing power under the dashboard of an automobile than was carried on board the first manned capsules. Our early astronauts relied upon huge computers at NASA control center, and a good communications system.

Like all children of my generation I followed the adventures of Buck Rogers in comic strips. On Saturday afternoons I'd watch a movie serial in rapture as Flash Gordon saved planets and aliens from Ming the Merciless, the evil Emperor of Planet Mongo. But I had no idea of the impact that rockets would have on my life.

To build hardware on the leading edge of new technology has given me a natural high. When anyone asks how I came to work on space programs, to erect underwater structures, or design computer equipment—the only possible answer is that I stumbled into the best career which anyone, especially a dreamer like me, could ask for.

1

"An aviator starts with a full bag of luck and an empty bag of experience. His goal is to fill up the bag of experience before the bag of luck is empty."

—Observation by an unknown pilot

The early days of space conquest were filled with the exultation of small successes, and frustrating failures. Space efforts were front-page news. Being an open society, our government published time schedules for any venture. The Soviets used that information to top any goal we attempted.

In late 1956, to coordinate the company's growing involvement with military and civilian rocket or satellite programs, General Electric organized their Missile and Space Department in Philadelphia. Offices were built on three floors of a warehouse at the corner of 32nd and Chestnut Street; the building was shared with its owner: Great Atlantic and Pacific Tea Company, better known as A&P grocers. At the time, I worked for GE's Naval Ordinance Department, Pittsfield, Massachusetts, and was determined to complete my college degree—somehow. With no school closer than Rensaleer Polytechnic Institute, fifty-six miles away in Troy, New York, I'd begun to wonder how that would ever happen.

When the new department began a search for personnel, rapid promotion was offered as an incentive to relocate. I seized the opportunity to work on programs that sounded dream-like, and looked into the details of a transfer. Upon being told that Drexel Institute of Technology was located across the street from the office my decision solidified like concrete. In June of 1957 I moved my family to Drexel Hill, a suburb of Philadelphia.

Timing could not have been better; orbit of the 184 pound Russian satellite, Sputnik-1, occurred on October 4, 1957, and stunned the world. The Soviets declared, "Today there is a new moon in the sky." To Americans, after a decade of Cold War and efforts to equal or overtake their space program, it was a bitter day. No one knew for sure whether the USSR had a massive lead in technology, but their rockets lifted huge payloads compared to ours.

Lyndon Johnson, then Senate majority leader, said, "Whoever holds the high ground of space will control the world."

John D. McCormack, a member of the House of Representatives Committee on Science and Astronautics proclaimed, "The United States faces national extinction if we do not overtake the Soviet Union."

Newspapers and magazines trumpeted those statements as a call for U.S. leaders to take drastic action. The Cold War expanded to include a Space Race. Hawks and doves in Congress took to flight. Until then, each military service had pursued development of its own variety of missile systems; collaboration between the branches was essentially non-existent. Together they acted out words that I remembered from an old novel, "He jumped to his charger and rode off madly in all directions."

President Eisenhower attempted to alleviate concerns of the country, and said, "Sputnik represents an achievement of first importance, but...as of today the overall military strength of the free world is greater than that of the Communist countries." Despite Ike's background as former supreme commander of allied forces during WW-II, his speech did little to assure our allies, or silence the critics.

A few days later, on November 3, 1957, as if to rub salt in our wound, the Soviets launched Sputnik 2. This time their satellite capsule contained the canine Laika as the first space passenger.

At Thanksgiving, my father and I had a chance to talk. He knew that I couldn't say much about work because of security. We spoke about common news, what everyone read or heard. Dad must have sensed that underneath a calm facade I was concerned. He settled a hand on my shoulder and told me, "This will work out in time, wait and see. Our government always seems to need some initial nudge to get its ass into gear, but they're damn good at playing catch-up."

The first U.S. launch after Sputnik was December 6, 1957. Vanguard, a Navy missile, rose six inches, settled back to the launch pad and exploded. One newscaster described the scene by saying, "It looked like a fat man collapsing into a Barcalounger." Morale of the American public reached a new low. Confidence in our capability eroded. The London Daily Express headline declared, "U.S. Calls It Kaputnik." At the end of 1957, even our allies felt the space race was one that the United States was losing.

◆　　◆　　◆

Missiles aren't a new phenomenon. Though little noticed until mid 20th century, they have crossed earth's sky for twenty-five hundred years. In 4th century BC, an Egyptian flew a steam-driven, bird-shaped object for pharaoh's amusement. Early in the 12th century the Chinese introduced fireworks, and colorful displays were thrust into the night sky by black powder. At Kai Keng, in 1232, Imperial troops drove off Mongol invaders by firing rocket-propelled arrows.

The roles of missiles, gunpowder, and military grew up together, but it was Europe that became reshaped by black powder—not China. After 476 AD there was no emperor or court in Rome. Every city-state, feudal lord, self-declared king, and pope used weapons that came to hand. Leonardo da Vinci was prominent among those who designed and cast artillery. His workbook shows cannon and siege mortars drawn in 1450. Missile-like shapes are also sketched there, but no explanation is offered as to how he might have used them.

British artillery used Congreve rockets that weighed one hundred pounds and had a three-mile range. In the 1812 attack on the new American capitol at Washington, D.C., noise and smoke confused the defenders. The city was abandoned, and our twelve-year-old presidential mansion was put to the torch before enemy troops withdrew. Rebuilt, the smoke-covered, scarred surfaces were painted white to hide the damage. That same year, Francis Scott Key viewed the American flag still flying over Fort McHenry during another engagement. To the dismay of sopranos at all future sporting events that scene inspired him to write the memorable phrase, "And the rocket's red glare," into our national anthem.

The last half of the nineteenth century found men of science holding discussions about space travel. Jules Verne wrote *From the Earth to the Moon* in which a group of three humans and two dogs were launched from a huge gun barrel formed by tunneling into the ground. Published in 1865, it foreshadowed the future: the vehicle was a conical projectile rather than the round cannon shot then common, and their launch was made from Florida. But Verne's travelers orbit the moon, never to return.

In 1895, a Russian schoolteacher named Konstantin Edouardovich Tsiolkovsky said, "The earth is the cradle of mankind, but one cannot live in the cradle forever." He showed further vision when he proposed the use of liquid hydrogen as fuel for a space vehicle. Herbert George Wells quit teaching that same year to become a full-time writer. *First Men on the Moon* is his tale of where humans met wild moon creatures, which he named Selenites. His voyagers managed a hair-raising, last minute escape to earth.

Georges Melies combined the stories of Verne and Wells into the first science-fiction film. Released in 1902, *A Trip to the Moon*, or *Le Voyage dans la Lune,* used the launch, capsule and space trip as seen by Verne, but the escape and landing portrayed by Wells. The concept of space travel reached an ever widening, international audience.

In 1903, the Wright brothers flew an airplane under its own power. Aloft for fifty-eight seconds at Kitty Hawk, it covered one hundred and twenty feet. Less distance than the wingspan of a B-52 bomber, but the tiny craft changed the world's travel mode forever. It took nine more years before another inventor's concern for the pilot evolved into the first successful parachute jump. Germany equipped their pilots with chutes in World War-I. The Allies did not.

Material handling limitations of that era made solid rocket fuels easier to use than liquid or gas. Robert Goddard, an American, experimented with heating radioactive material for thrust or propulsion. When the U.S. entered the First World War, he submitted his design for a solid fuel rocket to the U.S. War Department. It was taken under review, but later declined. Also in 1917, Forest Madson directed a film called *Heaven Ship*; it presented the first concept of a voyage to another planet—Mars. Now desolate, it once had a near earthlike atmosphere. Considering the transportation modes available on earth in that period, any concept of colonizing another world was radical.

Rocket societies organized in Europe, Russia, and the United States during the Roaring Twenties. Stimulated by writings of Alexei Tolstoy, Soviet film producers flirted with science fiction in the 1924 film, *Aelita,* where students were portrayed to lead a revolution among the downtrodden Martians.

Herman Oberth and Willie Ley were two renowned members of the German Rocket Society. A movie studio approached the group in 1928, for advice on hardware. Strapped for funds, the Society became film consultant to a comedic-drama called *The Woman in the Moon*, or *Die Frau im Mond*. Members designed and constructed realistic props. When the film was finished the society kept the hardware as their consulting fee.

Fritz Von Opel drove a rocket-propelled car at sixty miles an hour, and the Great Depression arrived in October of 1929 when the Dow Jones average plunged from 387 to 41. Entertainment industry's newspaper, *Variety*, ran the headline, "Wall Street Lays an Egg." Three thousand banks closed, and thirty million people had no income. Dad worked as a bridge painter for the City of New York, twenty-three bridges link Manhattan to the other boroughs. The hazard made it unlikely that too many people would compete for his job. My only sibling, Warren, had been born in 1928. I arrived in July of 1930.

Bridge of size. Not only is it a big job to paint New York's Manhattan Bridge, but it took two fotographers to get these shots of the work. The dizzy assignment fell to the lot of News cameramen Hoff and Meurer.

Painter Crews Rigging Scaffolds on Manhattan Bridge

Germany began to change noticeably in the mid-Thirties. Willie Ley came to the U.S. to write, and work as a propulsion consultant. Oberth started work for the Nazi's. In 1934, Adolph Hitler ordered destruction of the six-year-old rocket ship models, film negative, and all available prints of *The Woman in the Moon*. Resemblance between film director Fritz Lang's concept of the future, and prototype weapons being perfected by the Third Reich was too strong.

Now and then dad bought a pulp magazine such as *Magic Carpet*, or *Amazing Stories*, where I read work by Robert E. Howard, the author of fantasies such as Conan the Barbarian, and King Kull. It's hard to pick a favorite from the wealth of pulp writers who published their work in that medium: James Blish, Talbot Mundy, Sprague de Camp, and John D. MacDonald. At the public library I read fiction by H.G. Wells, Jules Verne, or Ray Bradbury, and learned of moon rockets, radar, time travel, and atomic submarines.

Radar was first developed in Britain as a defense tool, to look into northern France as a warning of bomber or fighter buildup; the initial installation became operational in 1940. Bird carcasses around the antenna area caused additional study of the waves emitted. It was the forerunner of cooking units known as radar-ranges, and later called microwave ovens.

The Brooklyn Dodgers were ahead 7—0 when, at 2:56 p.m. EST, Sunday, the Mutual Broadcasting System broke into the play-by-play description. "Ladies and Gentlemen: We interrupt this broadcast to bring you an important bulletin from the United Press. Flash, Washington: The White House announces a Japanese attack on Pearl Harbor. Stay tuned for further developments as they are received."

The depression had played out its worst by the time the Japanese shoved us into World War-II. Nineteen ships including six battleships either sunk or sustained damage. One hundred eighty-eight planes were destroyed, and one hundred fifty-nine damaged. Military and civilian deaths reached 2,403, and 1,178 were wounded. No more fence sitting. Ready or not, U.S. isolationism shattered. Within days the country was engaged on two fronts: Japan in the Pacific, Germany and Italy in Europe or Africa. My cousin, John Bishop, and half of his classmates, left Notre Dame University to enlist. He was stationed in California's Mojave Desert, and trained on B-24 aircraft; later, his squadron performed photoreconnaissance in the Pacific.

John's father, my uncle Milton, was superintendent of docks for the city. He assigned piers to ships entering the harbor; only a handful of them could accommodate the Queen Mary, Queen Elizabeth, Normandy and other passenger liners that had to be converted as troop carriers. Now it fell to him to coordinate protection of the piers between city watchmen and the military.

At age eleven, I pasted world maps on my bedroom wall to follow the progress of allied forces; it seemed to take forever before I could really see 'progress' on either front. In the spring I planted a victory garden, collected newspaper, scrap iron, toothpaste tubes for their lead, and meat fats for their glycerin. The family used ration coupons for scarce items such as shoes, meat, and sugar. Compared to how our family survived a lack of money during the depression, rationing was easy. Now we had money, but there was less to buy with it.

Grand Army Terminal, on the northwestern tip of Brooklyn, was the scene of ships loading with war materials: tanks, trucks, artillery, jeeps and anti-aircraft guns. Freighters and tankers remained in the harbor until a convoy formed. My family rode the 69th Street nickel ferry when we visited our grandparents on Staten Island. Anchored close, all of the waiting ships had to turn twice a day to keep their bow into the tide. Ferry pilots held a tight course while the other vessels took turns to pull up their stern mooring so they could swing around the bow anchor. Watching those maneuvers never failed to thrill me; it was like a dance, but choreography on a mammoth scale.

American troops fought at first with WW-1 weapons and ammunition. For basic training, soldiers used sticks as simulated rifles. When United States marines started an island-to-island return in the Pacific, fully half of the troops who waded ashore carried decades old bolt-action carbines.

During the war, Germany continued missile development under the Army's Special Ordnance Development Section at Peenemuende. Walter Dornberger was Director of DOSD, and twenty-five year old Werner Von Braun was his aide; at least that was the way it showed on their organization list. Dornberger, an artillery officer, was the liaison between rocket engineers and military interests. Von Braun was arrested once, by the SS, on a charge that he displayed too much interest in the non-military use of rockets. Because of his importance to their on-going weapons program he was released.

Contrary to common belief the earlier Vengeance Weapon-1, or V-1, was not rocket-propelled; it was a flying bomb designed by the Luftwaffe. A revolutionary ramjet engine drove the vehicle to 400 miles per hour for 150 miles. V-1 was nicknamed a buzz bomb because of the engine sound, as long as it buzzed you were safe. Unexploded V-1's were sent to the U.S. and copied. Twelve hundred units were manufactured for a final assault on the Japanese homeland, but never used.

V-2 rockets burned alcohol and liquid oxygen as fuel; they carried a one-ton warhead. A total of 3,745 were launched against English soil or the liberated coast of Europe, and resulted in 2,407 deaths. More deadly than a V-1, they gave no warning. If you heard the explosion you were a survivor.

Near the end of European hostilities, Von Braun led three hundred engineers, technicians, and their families toward western Germany to avoid capture by the USSR. An American Army unit took charge of the group. Goddard's scientific papers had been the foundation of German rocket development, and the captives were surprised to learn how few Americans had even heard of him. Their scientists persuaded the U.S. military to reenter the rocket factory zone before Soviet forces clamped tighter control. One hundred V-2 rockets were retrieved, together with dies, tools and spare parts. As spoils-of-war, the western allies held most of the scientists and technicians; the USSR held production facilities, machinists and assembly workers.

German engineers launched three V-2's while English and American observers evaluated the technology. September 1945 saw a group of one hundred twenty-seven depart for America, together with fifty tons of journals, drawings, and hardware; their families were to join them later. The war in Europe was only six months ended; to stave off protests from the American populace, Army officials didn't settle large numbers in any one location. Small groups dispersed to: Aberdeen Proving Ground, Maryland; Fort Bliss, Texas; White Sands Proving Ground, New Mexico—wherever Army ordnance was designed, built, and tested.

While little thought of as a rocket, the Army's bazooka was one of America's most effective WW-II weapons. It was portable, and swiftly moved into position among the infantry forces, unlike heavier artillery pieces. Soldiers were easily trained to its use, and it was deadly in its efficiency against tanks or armored vehicles.

The U.S. Navy was interested in missile science, but didn't request the presence of any German scientists or technicians. They wanted rocket-propelled weapons built to their own specifications; it was their opinion that only naval personnel knew how to provide the fail-safe handling needed for fleet operations.

Hiroshima was destroyed on August 6, 1945, by the first use of an atomic bomb. The Soviet Union chose to declare war against Japan on August 8th. Nagasaki was the second city to be bombed, on August 9th. For years, historians have debated the reason for Japan's surrender: the bomb, or the threat of Russian occupation.

When World War II ended, fifty-five million people had been killed, wounded, or displaced; thirty million were refugees. Cessation of fighting didn't end hostility. The Marshall Plan provided food or medicine, and the allies rehabilitated people in their homelands when possible. Many had to relocate in safer places. Military and political resistance by the Soviet Union besieged conquered countries as well as those in the western alliance. Short of outright warfare, the USSR opposed their former allies in every conceivable way. Western newspapers dubbed that period 'The Cold War.'

In a few decades jet engines and rockets have changed our planet; they promise to change the universe. A jet engine had been flown in 1939, but set aside for much of the war; it was 1943 before the first jet fighter engagement occurred. One problem was that an ejector seat had to be designed, to allow the pilot to get clear of a damaged aircraft. The ME-162 did have a basic ejector seat, but it was the British who designed a long lasting solution: a pilot's seat was literally attached to, and fired from, a small cannon. That method is still in use today. Another problem was that parachute fabric shredded under winds of more than three hundred miles per hour. Parachutes became usable for jets when a small, tough, drogue-chute was introduced, to slow a pilot before the main chute release.

The ME-163 Comet was Germany's first rocket fighter plane. Maneuvering was difficult at high speed, but the rocket craft was useful to fly through an Allied bomber group to break up their tight formation; then, their other fighters could engage smaller groups. The Luftwaffe had flown a rocket-propelled airplane to 596 miles per hour. Prototypes of the craft were shipped to America, and the Army Air Corps awarded a six million-dollar contract to Bell Aircraft Company for development of an advanced version.

Muroc Field is in the remote elevation of California's Mojave Desert where my cousin trained. Surrounded by sagebrush, dry lakebeds, and Joshua trees, it was an ideal area for testing of a new aircraft. The first experimental, supersonic, rocket plane was dubbed X-S-1. Delivered in 1946, it was kept in a hangar at the far end of the base. Shaped like a bullet for aerodynamics, but with a stubby twenty-eight foot wingspan and high tail, it was expected to be the means of breaking Mach-1—the speed of sound. Mach is a variable dependent upon wind speed, altitude and temperature. At sea level it is 750 miles per hour. At 40,000 feet with a temperature of sixty degrees below zero, Mach-1 is 660 miles per hour.

X-S-1 was loaded under a B-29 by backing into a cross-shaped pit and pulling the bomber over it. The test plane was shackled in the bomb bay of the mother craft. In case of an abort, the test pilot stayed inside the bomber during takeoff. After takeoff, the pilot climbed down a short ladder to reach the cockpit. At 25,000 feet the bomber pitched to a twenty-degree dive-angle. When the mother plane's speed reached 240 miles per hour, a countdown was given and the rocket plane dropped. To familiarize pilots with the feel of the aircraft and its controls, their first flights were static drops without the volatile fuel. But, dead-stick landings were always the mode; the engine shuts down when fuel is depleted.

Bell test pilot, Slick Goodlin, had flown the aircraft to 0.85 Mach, as required by his contract. He wanted a $150,000.00 bonus to proceed further into the unknown, where some fliers had tried and died. Army Air Corps decided to use their own test pilots. The aircraft designation was changed; it was shortened to X-1.

David Clark, a Massachusetts corset manufacturer, received the first contract to make high-altitude flight suits. Early suit models were stiff and hampered movement. Yeager and Ridley, the first pilots sent for fittings, joked about looking like Pillsbury doughboys. They also expressed concerns about whether a pilot could manage to eject if necessary, but it was all that was available. Hard hats were unknown then, the leather helmet and goggles used through both World Wars remained standard issue. Buffeting of the X-1 during static drops had made Yeager aware that he could be knocked out or dazed during severe movements of the aircraft. He cut ear holes in a tank-driver's padded leather helmet, and wore it over his soft headgear.

On October 14, 1947, in an X-1 vehicle on which the pilot had painted 'Glamorous Glennis,' in reference to his wife, Chuck Yeager became the first human to break the invisible sound barrier. For that spectacular achievement he earned $783.00 per month plus living and uniform allowances, standard pay for a captain in the Army Air Corps. Over the next six months Captain Yeager flew supersonic a dozen times—also unannounced.

Based on unconfirmed rumors the December issue of Aviation Week contained a story claiming that the X-1 had surpassed Mach-1, but it didn't name the pilot. Yeager's feat and identity were not confirmed until June 1948, after the Army Air Corps was reorganized to become the United States Air Force. Muroc Field was expanded and renamed Edwards Air Force base, but Yeager's story was pushed off the front pages by other events as the Soviet Union hurled yet another surprise at its former allies.

Germany had been divided into zones; the capitol, Berlin, was inside the Soviet zone. Their border guards provoked delays for the convoys that delivered supplies through the Berlin corridor. Vehicle traffic ground to a standstill. Residents of the city faced starvation and freezing. The United States refused to be bullied into surrender of the capitol and created an overhead route known as the Berlin airlift, which continued for nearly two years before the Soviets stopped their harassment.

Author
Yearbook Graduation Photo
Fall 1947

I graduated from Brooklyn Technical High School in January of 1948, and my original goal was to teach shop and drafting at a vocational-technical school. For that, I intended to earn an Industrial Arts degree. To gain experience while saving for tuition, I left home to enter an apprentice program with General Electric's Transformer Division in Pittsfield, Massachusetts. Later I married, and postponed my college plans.

Postwar sci-fi films were dominated by fears of nuclear fission. Director Burt I. Gordon, called Mr. Big due to his initials, made B-grade flicks with a theme of radiation mutants. Movies such as *The Giant Crab*, *Them*, and *Attack of the Fifty-Foot Woman* attracted large crowds to the drive-in theaters. Ironically, Japan, the first to feel the effect of atomics, created a series of radiation mutant monsters. *Godzilla* and *Mothra* began a string of look-alike films that became one of their most profitable postwar exports.

On June 25, 1950, North Korea invaded the South. American presence in the region was limited to a few thousand advisors for the South Korean Army, which was being trained to act more as a police force than a military unit. Headquarters for Douglas MacArthur, General of the Pacific Theater, was located in Japan. The bulk our forces with combat experience had been discharged. U.S. Army Reserves were ordered to ready-status; World War-II navigators and pilots were recalled. President Harry S. Truman, nicknamed Give 'Em Hell Harry, formulated an Asian support program known as the Truman Doctrine. Alone, America aided South Korea against communist aggression until the United Nations could be rallied. I was enrolled for the draft, and returned draft board questionnaires, but with my spouse expecting our first child I was never contacted again.

Early years of the Fifties saw the return of space travel as a movie theme. *Rocket Ship-XM* brought aliens to earth. That was followed by *Destination Moon, Flight to Mars,* and *When Worlds Collide.* Later in the decade came *Forbidden Planet,* and *The Day the Earth Stood Still.*

The USSR exploded their first nuclear bomb in 1953, and western nations were alerted to still another means of how the Soviet Union could intimidate. For decades there was concern as the military on both sides, east and west, built, and stored massive weapons of destruction.

On November 20, 1953, Scott Crossfield flew a rocket-aircraft at Mach 2.0. A month later, Mach 2.4 by Yeager, and the X-aircraft programs continued. In the latter part of that year Crossfield piloted an X-15 to 280,000 feet. Slightly over fifty miles high, it was then regarded as the boundary of space, where atmosphere is hardly detectable.

About the same time as GE formed their Missile and Space Department in Philadelphia, late 1956, Wright Air Development Center in

Dayton, Ohio, launched an Air Force study of manned space vehicles. It was simply called Project 7969.

◆   ◆   ◆

By the time I relocated, in mid-1957, work experience in industry had convinced me to change my baccalaureate goal from teaching to engineering. However, there were times that I doubted whether I'd ever become an engineer, or finish school at all. The programs I designed hardware for were nothing short of fantastic, and the original department grew to a division with three departments. My job level and salary improved as time passed, but—with regard to college—so many things went wrong.

2

"The Impossible Dream,"

—Theme Song: Man of La Mancha.

When I transferred to Philadelphia I'd considered my academic problems solved, but Pittsfield was the only GE location where apprentice class records weren't kept by a state office of Higher Education, or a Regent controlled organization. Massachusetts provided classrooms and instructors, but the company maintained our records. I had taken fourteen classes that no college would accept for academic credit.

In the fall of 1957, I started an engineering program at Drexel Institute. Today it's known as Drexel University. The following April, President Dwight D. Eisenhower reorganized NACA, the National Committee for Aeronautics, to include space. NASA was born, and GE's business exploded. Hours of overtime were added for new studies and proposals that were performed by the design group in which I worked. Overtime money was welcome, but the hours were long, and when extensive travel was required for a job I had to drop out of class. Progress toward my baccalaureate moved at a crawl.

A grandfather clause in the engineering law allowed experience and limited classwork to qualify applicants for state examinations. GE employed those who held either a license, or a degree. I applied, passed, and my license to practice as a professional engineer was issued in 1963. Next time a position opened up in my office I expected a promotion, but rapid corporate growth had begun to slow. For two years, there were no openings.

Al Mitnick and I were friends as well as coworkers. His hair dipped over his forehead in the manner of Buddy Holly. He had a contagious smile and laid-back manner. We had worked together five years, and he'd recently become the manager over my design group in Advanced Engineering. He was getting married, and needed help in moving from an apartment to a house. We talked as the U-haul van was loaded.

"What are you doing to advance your career?" Al asked. "Don't waste your license after all of the effort to get it."

I was surprised by his comment; to me it seemed obvious. "I'm waiting for a mechanical engineering position to open up in our office."

"Well, the basic requirement to work in an advanced group like ours is to be in the top of your field—top ten percent—the best. As an engineering-designer you're in the right spot, but you don't have the experience to be an engineer. Not yet. You should start a search through the Personnel office."

There it was. What I'd waited for wasn't going to happen unless I moved to a different group. "Where can I get good experience, Al? Maybe I can return later."

"Don't get too serious on me. You'll find excellent work in other groups. Fill out a transfer request. I'll sign it, and that gives you permission to look."

"Then what?"

"They'll start a job search, and look for someone who needs a mechanical type. We'll review what they come up with."

"In the meanwhile?"

"It's just permission to look." Al grinned. "No one is telling you to leave. Stay with Advanced Engineering until something fits the way you work. It may take time. Probably be better if it's in a small unit like ours."

My face must have told Al that I didn't understand what he was saying.

"You're a practical, hands-on type. Doing stress-analysis at a desk all day would bore the hell out of you. Wait for a job you'll like."

Personnel had two requests on hand, and they arranged interviews. The first was a missile program; the other was a communication satellite. I told those managers that I'd think about it. There were four interviews that week, but nothing seemed to be a good fit.

Monday morning, Personnel called. "Can you meet with Fred Parker, today?"

"Sure, but who is he? What group is he with?"

"A simulation group—part of MOL."

Personnel people loved to use acronyms. I asked, "What's that?"

"Manned Orbiting Laboratory—an Air Force job. Parker needs a mechanical engineer on board—like yesterday. How about one o'clock this afternoon to meet with him?"

"One is fine. What room number?"

His office was at the opposite end of the half-mile long building; the cafeteria was about midway. After lunch I turned left instead of right, arrived on time, and knocked.

"Don't be too formal," I heard. "Come on in."

The room was cramped. I saw a typical manager's eight-by-twelve-foot office, but it was crammed with hardware and books; Fred was more concerned with work than appearances. I looked him over; he was short, and looked like he was pouting. Then I realized that his lower lip jutted out and he had a habit of flexing it. That made it look larger. His receding hairline was every bit as bad as mine.

"Good to see you're prompt, I need to be at the airport by three-thirty. Do you scuba, or at least swim?"

Personnel hadn't mentioned anything about water. "Swim, yes, but scuba? No. Is that a requirement?"

He sighed. "No, but it would help. At least all of you swim. So far no one has done any scuba except me. Sit down."

"Thanks," I replied, and squeezed into a chair turned sideways. In fact, it was trapped in that position by the desk, the wall, and a book-case.

"Now, what you came to hear. My goal is to find people who can get things done without having their hands held. Most of us will be traveling a lot."

Underwater Test Facility,
Little Buck Island, USVI,

Fred pointed to a folder on his desk. "Your last review indicates you work independently, without a lot of support. It also says when you hit a snag, you know who to ask for help and keep going."

He located two eight by ten glossy photographs from a pile on his bookcase. An artist had captured Fred's vision of the program. The first photo was a view from above, to show the island cove, a submerged structure, and an observation tower in the water. On the shoreline was a flat roofed building.

The second picture showed a missile like structure under the water. But it was a skeletal shape composed of rings, lengthwise braces, and covered with a wire-mesh skin. Test equipment was packaged in the forward part of the structure, and scuba divers were approaching the airlock.

Fred leaned forward and closed the gap between us. "Nothing in that bay exists today. If you join us, you'll create them before the end of this year. There isn't a tank or pool facility large enough to hold the structure we're planning. Capital expenditure or long lead times rule that out at this time. We'll do the program at Little Buck Island, it's five miles south of St. Thomas in the U.S. Virgin Islands." He leaned back. "Is travel a problem?"

"Hasn't been so far," I replied.

"I mean lots of travel. Building the structure and outfitting Buck Island is only the start. Testing and training come next. We'll be on site for weeks at a time. The simulator will be an SIV-B, the top stage of a Saturn-V moon rocket, with an airlock that Lockheed has just designed for a project called Skylab."

SIV-B Simulator for the Orbital Workshop Program, Renamed Skylab

"What work is there to do, once the site is operational?"

"Well, for long term travel to Mars or Venus, astronauts can't sit in a capsule for a year or more. NASA wants to use Apollo on top of an SIV-B stage. If it's possible to reseal that tank when the fuel is expended, astronauts could bring their equipment from storage and set up living quarters inside the last stage. They want answers; we want to provide them.

"Eight months isn't much time," I said, half aloud.

Fred sensed that he'd roused interest. "We don't have to do everything at once. If we get things in place by Thanksgiving, and do any testing in December, we'll have next year's budget. Leave that in Carl's hands; he's the money man." He briefed me on other aspects of the program. "GE is funding this in expectation of a payback, a contract for underwater services. And NASA wants to evaluate our method of underwater life support versus theirs, to see if ours works better. I consider mine safer, but we'll need to convince them."

"What's the difference between the two methods?" I asked.

"NASA uses air through the umbilical connection to inflate the suit. For neutral buoyancy in water, a minimum of one hundred and fifty pounds of weight is needed: belts around the arms and legs, plus crossed bandoleers around the body. My rebreather is self-contained, no air hose, and the suit is pressurized with water instead of air. I've tested in a Gemini suit with fourteen pounds ballast. And, no tether—I can go anywhere."

To be hands-on in a project such as Fred described sounded incredible, and the job was mine if I wanted. My head reeled with contradiction. Not much time. Do it and next years funds were assured; don't do it and you'll be back to Personnel. I entered this room to find something that would aid my career as an engineer, not to look for adventure. Two thoughts were in conflict. "What a challenge," was countered by, "You've lost your mind."

Fred reeled me in. "Do you want to design the equipment so it can be brought to the surface for repairs, or learn scuba and supervise anything that needs attention down below?"

Without consciously willing it, I heard my own voice reply. "I'll learn scuba."

"Hoped you'd say that. My equipment was bought from Camden Marine, across the Ben Franklin Bridge in New Jersey. Get everyone set up with scuba gear. Organize a class for ten people. Find a pool where we can train, choose a time—maybe after work—plus an instructor. Then post training times. We can't all be there every time but we'll have to work it into our schedules. Plan for a few make-up sessions also."

"Who will handle our in-house construction?" I asked.

"Rocky Caruso. His shop is on the next aisle. He has a four-foot high swimming pool set up, to allow me to test new features for the rebreather. I'm using a Gemini suit now, but we'll have Apollo suits whenever the astronauts scuff them up."

"I've worked with Rocky before," I said. They built my models and mockups for several projects with Advanced Engineering. They're a good group, easy to work with."

"Well," Fred added, "his people will make lots of hardware for us, but I doubt that any in-house shop has capacity for the large structural parts of the simulator itself. Give Rock the drawings when you have 'em, and he'll get bids. They should be finished with their current job in three weeks."

"That's as soon as I could expect to have any drawings. The simulator will have the most piece parts to it. I'll start there first."

"There are two empty desks next door. Pick one and settle in."

"Who else is working there—next door?"

"Jack Burt, our physiologist, but he's in class today; he'll be here tomorrow. Charlie Soult is our mechanical designer, he's there every day—right now he's drawing new rebreather parts." Fred handed me a folder with the photographs and SIV-B drawings. "Work up ideas on how to simulate that shell. I'll see you Thursday."

If it sounds abrupt, it was. I was hired, given my orders, and turned loose in an hour. Later I found out that Fred wasn't being rude, it was his modus operandi. He told you what to do, defined assets and authority you have, then didn't expect to hear from you for a while.

I went to the office next door. Two rows of desks faced each side of the entry door, and two bookcases occupied half of the center aisle; there wasn't room for anything else. The desks were so close that it would be necessary to move the swivel chairs into the middle aisle to even turn them around. Charlie sat at one of the front desks. When I told him I'd be moving in, he launched into a briefing on the air regulator. As the lead mechanical engineer I'd collaborate with him, but I asked him to wait on that.

"Do you mind if I suggest how we could have more usable space in here?"

"Not at all, anything would be an improvement."

"Let's turn the two rear desks to face the back wall, so that all of the open space will be in the middle. That will make it easier to sit and work. When anyone is on the road, it will be better for whoever is left in the office." We did that, and put each bookcase against a side wall. Now there was room to maneuver the chairs.

"By the way," I asked, "do you have a drawing board around here?"

"No. Rocky has a four-foot board in his lab. There's no way we could squeeze one in here, even rearranged like this. Why?"

"I'll be starting the simulator layout, a large one. We'll probably have several hundred piece-part drawings to make. I need to give some thought to how we can best do that." I rose to leave. "See you tomorrow."

I stopped at Personnel on my way back, and gave them my verbal acceptance. The clerk asked when I'd report to the new assignment, but I needed to allow Al Mitnick the courtesy of saying when.

The pictures and a description of our task captured Al's interest from the first word. "You see what I meant about the right spot?" he said. "It sounds fantastic, and I see the excitement in your face. You talked money?"

"Sure did. My salary will be the middle of the next pay level. When Fred talked about scuba, and the Virgin Islands, I wondered if I'd have to pay the company instead of them paying me."

Al laughed, and told me to pack. "I'll get your gear moved for you tomorrow." We shook hands, and he added, "Don't forget to send postcards."

That was a perfect example of synchronicity. Everything brought the event to me; it was impossible to avoid. Later, when I told Dad, he said, "You're the one who inherited the pure dumb luck of the family."

Being in Advanced Engineering for seven years I'd accumulated tools and books that would never fit my next office. I divided it: one pile for the new room, one to disperse among friends, and one to take home. The latter pile was the largest, but that was appropriate; I was changing careers, and had to revise my image. I'd make tons more sketches in my career, but there is a definite difference in the way management views a draftsman making drawings and an engineer making sketches. But I did have to give thought to how we'd make our drawings. To spend time on a board in Rocky's lab would remove me from group interaction. A phone call to Facilities told me there were boards in storage; they'd check sizes and let me know. Two desktop units were located—I had them sent to Charlie Soult.

Drafting service area was at the center of the building; they ran all of our drawing prints. I'd need copies for vendor bids, and four hours was the average turnaround. A copier machine stood fifty feet from the new office. If drawings could be run there it would be fast, but it would limit me to eleven by seventeen-inch drawings, a B-size drawing in drafting terms. I checked an office supply catalog, found that they carried pads of semi-transparent graph paper in that size, and ordered a case. An exception to B-size drawings was the SIV-B simulator assembly, but I was the only one who would be inconvenienced waiting for copies of that. For the rest of the day, I reviewed NASA's outline drawings of the rocket portion to be simulated. Blueprints provided the location of one hundred and twelve attachment points for internal equipment. They provided realism, and that was important to our planned duplication of astronaut tasks.

Twenty-two feet in diameter and, I estimated, nearly sixty-feet of the adapter and fuel tank would be required. Standard tracing paper comes in rolls thirty-six inches wide. If I were to draw it in one-tenth scale the structure would be about twenty-six inches high and seventy-two inches long on my assembly drawing. That was just the side profile, without any cross-sections. Several of those were needed to show details of the interior equipment and attachment hardware.

Next morning I went directly to my new office, and began to search the yellow pages for groups that taught scuba. The YMCA didn't offer any scuba training for six more weeks, and that was a once-a-week class. It would take too long for us to prepare. They wouldn't teach a separate group; that would interfere with their daily schedule of water-aerobics, lap swimming and free swim. The place I'd expected to be most helpful didn't pan out.

I scoured the phone book, and called anyone listed under pools, pool cleaners, or pool equipment suppliers. None worked out. The search continued on and off for the next few weeks, but design of the simulator had to be started. My bookcase joined the one across the room and the empty wall became my drawing board. With an eight-foot length of paper taped in place, I drew reference lines with a straight edge and two triangles. Soon, the SIV-B profile appeared.

At first I drew only the outer shell. Then the attachment points were added. A tapered section in front of the fuel tank was originally built to hold the Lunar Excursion Module, or LEM, on moon landings. For SKYLAB, Lockheed had constructed a four-foot diameter airlock tunnel to fit that space. Four trusses anchored the tunnel in the center of the outer shell. Reported in *Aviation Week*, the story was complete with photographs. Having no other information source, I measured the magazine pictures, prorated the size of the important parts, and placed more information on my composite drawing. Details were added to that layout whenever an item was discovered that I felt should be included.

Fred returned on Thursday, and there was plenty to discuss. We wrote on my rough layout to express new ideas. The structure would lie on its side; I sketched a cradle under the shell, to steady it on the sandy bottom. Fred felt that a trainee in a spacesuit would require help to move around. For safety, in case of a need to escape the cylinder, emergency hatches were added to the top section. They would provide a fast way to break out and reach the water surface.

My eggshell-like structure now met all requirements. I divided the length into ten similar but not identical sections. Twenty-two foot rings were to be rolled using two by three-inch aluminum angles with one-quarter inch thick cross section; each had to be made in eight arc-shaped pieces, to cover the seventy-foot circumference. Ring strength was improved by rotating every other cylinder so the next cylinder's ring overlapped those segment joints halfway. I estimated that the simulator would weigh about seventy-five hundred pounds.

Fred's intention for the group had been one mechanical engineer, one physiologist, mechanical and electrical designers, a detailer, mechanical and electrical technicians, plus two test subjects—a team of nine. Only part of his personnel search was ever fulfilled: one mechanical engineer, one designer, and one physiologist. We could borrow other support for short times whenever it was needed, but it was beginning to look like a busy eight months. The group also had the limited services of two former Air Force test pilots who would assist as test subjects. They'd become more involved when the testing phase was reached. We found other ways to fill personnel gaps, and made use of Rocky's technicians. Bob Plank had been a detailer-draftsman elsewhere. Art Rachild, Eddie Sienko, and other technicians covered my electrical information needs. They were included in our scuba sessions, and on numerous occasions they worked as grunt labor at St. Thomas and Little Buck—both in and out of the water.

Sheer size of the simulator structure, the number of piece parts to be drawn and built, made it essential to be detailed first. Once I'd resolved the angle sizes to be used for the rings, and determined how the rings were to be formed, or cut into segments, then the final assembly outline was drawn. The three cone-shaped sections at the front included the airlock, and seven cylindrical units followed that. Although not identical, there were common features: longitudinal stiffeners, ring angles and safety doors were alike; I told Bob Plank to draw one picture wherever he could, and tabulate the dimensions of all parts that looked the same. He moved to the empty desk next to mine and began to detail the pieces of my assembly.

Meanwhile, I'd continued my search for a place to do scuba training. There were several GE employees who offered use of their home pools, but I felt that we needed an indoor facility where practice could continue when the weather turned cool. Naval Reserve office suggested scuba instructors, but didn't know of any pools. Weeks passed before I found the perfect place by accident. After lunch, I was passing the Valley Forge Motor Lodge at the foot of the hill the GE buildings are located on. I'm not sure what prompted me. On a whim, I pulled in. The manager was tending the front desk. I described our need, and asked whether GE could rent their swimming pool for training.

"We host a lot of your visitors here," he said. "Maybe we can help you out. When do you need it?"

"Late afternoon or early evening, two or preferably three days a week."

"That shouldn't be a problem," he said. "We don't have much usage during the day or through dinner hour. You'll have to be out no later than six-thirty though. Guests want to swim at night. Work up a schedule and I'll look it over."

"I'll need to write a purchase order. What will be the rental fee?"

He grinned. "Clean the pool while you're in there."

I stared at him.

"Sponge down the walls and floor while you're in there," he replied. "My maintenance man is going to love this when I tell him. He's been threatening to quit."

Once the motel agreed to let us use the pool, my next step was to find a certified diving instructor. Fred had talked earlier with a professional diver, David Stith; the staff at Camden Marine had referred him. He was a freelance diving contractor working out of his home in Cherry Hill, New Jersey.

Fred and I talked regarding use of Dave's services. Neither of us had met the man yet, but his credentials made him sound like the dive master we needed at St. Thomas. Someone had to oversee safety of the operation; all but Fred were rank amateurs at diving. I called Dave, and invited him to meet with us. A former Navy diver for both hardhat and scuba, he was a bulky figure of a man who exuded confidence. His white-toothed smile contrasted with the perpetual five-o'clock shadow across his chin.

Fred described what our program was to do. He asked Dave to contract with GE as a consultant when need be, and dive master at the Buck Island site. He'd be free to accept other work that didn't interfere with his availability for our program. I proposed that he could begin by teaching scuba to our group. Dave agreed. He felt that doing the scuba class would give him the opportunity to evaluate the team; if there were a problem, it would be wise to know it at the start. I worked up a late-afternoon three days per week schedule, verified it with the motel manager, and posted it in the office.

Equipment brought me back to Camden Marine, and I arranged for team members to go there at their convenience. Each selected a face-mask, snorkel, flippers, backpack, air regulator, and wet suit jacket. Most items seemed to be of modest price except for the air regulators; those are what made their inventor, Jacques Costeau, a wealthy man. I rented compressed air tanks for our practice sessions, with refills to come on a weekly basis.

Training was from three to six p.m. on Monday, Wednesday, and Friday. With work and travel schedules, I hoped that everyone would make two meetings a week. We were skin divers before proceeding to scuba; Dave wanted to evaluate everyone before letting us rely on neoprene rubber wet suits to keep afloat. Swimsuits, masks, and snorkels were used the first week. None of the group was an Olympic quality swimmer, but no one was likely to get into trouble.

Dave introduced every item with a demonstration. Facemasks: clean it of any film coating before donning. Spit in it and slosh it in the water to rinse it off. How tight should the strap be; I was surprised how loose it could fit and still hold its place. To clear it if water gets in, hold the mask at the top, snort through your nose, and the water goes out the bottom. Don't blow all the water out. Keep a tiny amount inside so you can slosh it around the interior glass surface to clear fogging.

The snorkel was added. Clench it gently between the teeth. Don't bite into it, and keep your tongue in the hole to hold water out when you dive. Dave had the opinion of most of the Navy people whom I've met, "Water is okay to drink, but there are better things known to mankind."

We donned flippers. Don't bend the knee. Keep it stiff and pivot the leg from the hip. No one shrank from our obligation to the motel management. Every time we used the pool there was a sponge in each hand, scrubbing the tiles. Dave kept us working in the shallow end until we learned to take a breath and dive lower. Snorkel breathing soon became as natural to us as breathing in air.

The time arrived for my first visit to St. Thomas and Little Buck; I needed to meet with local subcontractors we'd use. If there was a logistics problem that I could foresee, there was still time to adjust design of the simulator or the operations building. Fred and I flew to San Juan, Puerto Rico and changed planes. The terminal in St. Thomas was Harry Truman Airport. It was a metal-roofed shed with an A-frame like roof, with the tall ends open. Huge, noisy fans were mounted at one end of the building, to push air through the crowded ticket and baggage areas.

Al White met us; he was to be our interface on the island. GE couldn't hire workers directly; our state Workman's Compensation wouldn't cover them. Al owned a twenty-foot Mercury MerCruiser that we'd use for day-to-day operations. He would hire other boats, divers, and workers, as they were needed. An advertising man in New York City, with a hobby of sport swimming and scuba diving, he'd moved to St. Thomas five-years earlier. His employer had refused him vacation time to swim events in the Pan American Games, a tournament Al had won in the past. He gave notice, packed his wife and two children on their boat, and headed south. It's no mistake that his boat is named *Distant Drummer*.

Fred and I met with a St. Thomas construction company to discuss the twelve by twenty-four-foot building and boat dock. We agreed to erect a level platform on the old pier. Then the boat dock and building would be barged to Little Buck Island in prefabricated sections.

Later, they dropped me at West Indian Company. Fred had spoken with their dockmaster before, but I introduced myself to Lars Pederson. We discussed the type and size of area needed for the simulator structure. He showed me an open space behind the entry gate, away from the area where cruise liners docked. The surface along the edge of the pier was covered by individual cement pads that had shifted over the years; no two adjacent pads were quite level. But the area was clear of the storage buildings, and offered a feature that we'd been concerned about: security—it was within sight of the watchman's shack at the front gate. I agreed that he should store our cargo there. He asked me to walk with him to the office, to sign a contract.

We passed through a gate in a six-foot chain link fence. Once on the other side, Pederson said, "Someone may ask to see your passport when you want to return. Have you been outside the U.S. before?"

"No," I replied, puzzled by his question. "They told me I don't need a passport here. The islands are a territory of the U.S."

"Well, you just left the United States," Pederson answered. He chuckled. "You're on my home soil now." Then he explained that West Indian's offices are in the Danish embassy, four acres that adjoin the dock facilities.

When I finished, a short taxi ride brought me to Sebastian's "On the Waterfront Café," where Fred and Al would join me. Surrounded by ocean water, I noticed that the air had a salty tang, which gave me a strong desire for something to drink. I sat on the second story balcony with a Pina Colada in hand, and surveyed the street as far as Yacht Haven to the east. Small boats pulled in, and merchants put blankets down on the concrete quay. That open-air market had fruit, fish and vegetables on display—a mile long supermarket.

After lunch we headed toward Little Buck Island. I was excited, and anxious for my first look at what I'd mentally envisioned to be an island paradise. It had been selected as the project site because it was thought to be sufficiently remote from St. Thomas to discourage curiosity, but not so far as to cause operational problems. There were machine shop facilities, boats, divers and diver support shops five miles away. The cove is seven hundred feet in diameter, protected by the easterly trade winds, and only an eight-inch tidal rise. I'd read all of those details in Fred's notes. Also important is the year-round water temperature range of 78 to 86 degrees Fahrenheit. Even at those seemingly comfortable temperatures a body can lose significant heat when immersed for, say, three to four hours of work.

A billboard-sized sign on a high part of the island announced to would-be visitors that this was U.S. Navy property—off limits. The island was little more than stony crags, sparse grass, and weeds, with no bushes or trees to be seen. An automated lighthouse rose from the highest point. Water depth inside the cove was less than fifty feet, and I needed to inspect two sandy bottom patches as potential work sites. When the boat got closer to the inner shore, I saw the twin concrete strips of the one-time pier.

From the beginning I'd known that Little Buck Island had been a practice area for SEAL teams, but I never spent much time thinking about what that meant. The dock must have been their favorite demolition target. Its twin strips weren't level or parallel, not any more. Height or width in some places appeared to be two feet askew—in opposite directions. There wasn't a trace of the boards that had once covered the concrete. The dock where I had to erect a platform for the contractor was a disaster.

3

"The time has come, the walrus said, to talk of many things...."
—*Alice's Adventures in Wonderland,* Lewis Carroll

Construction of an operations building at Little Buck was key to using the island. Fred had decided we'd use the dock; it was up to me to see that it got done. I measured the shattered remains of the twin strips. Simple supports previously described to me gained in complexity. The worst height difference was twenty-eight inches. One strip tilted to the outside by eighteen inches. Another portion of it leaned twenty-four in the opposite direction for a total of forty-two inches. It would take a considerable range of adjustment to accommodate that, but better to have a hundred extra holes drilled at King of Prussia than for one to be missing when I'd come back with a crew to install those supports.

Once the dock measurements were complete, I wanted to identify the site where I'd place the underwater simulator. We'd changed to swimsuits after lunch, and carried our facemask plus the snorkel. Al showed me the two potential sandy locations for my simulator. With the SIV-B simulator dimensions now known, it was obvious that the sandy patch nearest the old dock was marginal. Divers outside of the structure would have to beware of sea urchin spines crowding that area; their barbed tips break off, leaving a hook to cause infection. The other site was further from shore, but larger and level.

It was the first of countless times I swam in that bay. My experience with east coast USA beaches hadn't prepared me for waters this warm; at eighteen degrees north of the Equator, and summer just starting. The water was about eighty degrees. After I looked over the potential sites, there was time to play tourist. I couldn't identify all of the colorful fish life in the cove, but it was abundant—another living world. Beautiful. Closer to the pier I saw brain coral, antler coral, sea fans and various shells. My daughter is a rock hound; but I knew that Sheryl didn't have any specimens like those in her fledgling collection. I didn't break anything off, but picked up five small samples that were loose.

Events of that trip made me aware how Fred's mind worked. He had clever ideas and would outline a desire, but once he'd sketched a thought his mind moved on. It was up to someone else to fill in details, and color in the mental outline he'd drawn. If he told me that a contractor was to do a job, the man might say, "Well, he did mention that," or, "He asked about that some time back. Is that what you want me to do?" Fred was a visionary, but once he described his desires it was up to me to pull the loose ends together.

When we returned to the office in King of Prussia, I checked Bob's drawings of the simulator piece parts. He made corrections for me, and I gave three sets of blueprints to Rocky. Fred's intuition had been correct. The structural pieces were larger than could be fabricated in-house. Rocky sent the drawings through the Purchasing office for quotes. Next, I designed a set of aluminum deck supports from structural channels. They had to be easily handled without a crane, and made in small sections that could be brought out on Al White's Mercruiser. Those pieces would be bolted together when we reached Little Buck.

The vertical legs were made adjustable for thirty-two inches in height. To allow for sideways misalignment and sloping of those concrete strips, the upright leg portion was attached to a flat plate by means of a ball-type swivel joint. That footplate would be anchored to the concrete with bolts into lead compression anchors. Although not all of the horizontal beams would require four and a half feet for width adjustment, they were made alike. To avoid any confusion at assembly time, it was best to keep the parts common and simple.

My design provided for a boat dock at the deep-water end of the concrete strips, and a four-foot wide staircase from the boat dock to the platform level. The building was set back six feet, for an observation deck above the dock. Platform supports were made wide enough to include a three-foot walkway parallel to the building, but now everything was larger than I'd discussed with the contractor. I made a final drawing of the boat dock, staircase, and railings. The vendor needed to review the changes and revise his total cost. Bob drew piece parts for the deck supports. This time, they were small enough to be made in the GE machine shop.

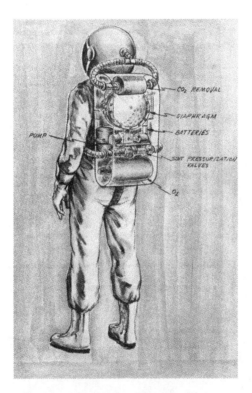

Astronaut Life Support Backpack, Concept

Fred's rebreather had to be repackaged, and I incorporated a water pump. My first design copied the Life Support System that feeds air to an astronaut, and pressurizes the space suit. I had to create a second unit to represent the EVA pack, or Extra Vehicular Activity pack, which is used on space walks. Over the period of time those underwater programs continued I repackaged the external shape of the rebreather five times.

Passing a test subject's exhaled air through barolime filters and purifies it. The carbon dioxide is absorbed, and a replacement fraction of oxygen is released from a tank to bring the breathing air back to normal composition. I used a tiny oxygen bottle from a fireman's rescue pack. For his purpose the bottle would last twenty minutes; for mine, it lasted eight hours. The barolime had to be housed in an easily removed canister that could be placed in an oven. Heating removes the moisture, and restores the material to its original condition.

Underwater Life Support Backpack and Helmet, Actual Hardware

In order to measure the physical well being of a test subject, another part of the pack housed: medical devices, recorder, and batteries. Heart rate, blood pressure, and stress levels couldn't exceed specified values or NASA would have to modify that task. The battery pack for our body function recorder was also readily accessible, so it could be replaced every five hours. On two occasions a battery was changed while underwater, in order to work through a task of eight hours duration. Most tasks were shorter than that, and repeated until the subject could do them without error.

Styrofoam blocks were inserted in and around the pack, to neutralize the weight in water. Later, Frank Pugliese, the project photographer, also found a use for foam. His twelve hundred-dollar camera housing had maneuverable diving planes on both sides. That was great if he was moving with a test subject, but when standing in one place he had to support and balance the camera and housing weight. That's tiring. He taped foam blocks on the exterior, to compensate for its dead weight.

Initial tests of the rebreather were conducted in the twenty-foot diameter by four-foot high pool at Rocky's lab. After the first few times, during which we gathered a substantial crowd of onlookers, we decided to work after five o'clock. Fred would put on the practice Gemini suit, but not the helmet. Then he'd climb over a ladder arrangement that Rocky's people had built. It had stairs from the floor up to a platform that spanned the side rail, and more stairs down into the pool. Someone beside myself must have lived in New England where a stile over a fence rail is a common sight.

Jack Burt, our physiologist, was usually the one who entered the unheated pool to help Fred with the rest of his dress procedure. After placing Fred's pack, and mounting the helmet, the spacesuit could be pressurized to 3.5 pounds per square inch. The pool wasn't tall enough to allow testing while he stood. To be completely covered by water, Fred had to lay down on the bottom—awkward—but it was the only means we had to check the system when the suit and pack were fully immersed in water. Communication methods were tried. A 'bone-knocker' works by vibrating against the skull near the ear, but it was a one-way system. Fred could hear our instructions, but couldn't respond; he wore a scuba diver's type of mouthpiece. Both the suit and helmet were filled with water.

Sooner or later we had to find a place to practice where Fred could stand up and move about. The Valley Forge Motor Lodge was still available until six-thirty at night, but workdays stretched longer than that. I'm not sure who brought up the idea of calling the Aquarium, but it turned out to be an ideal location for standup practice.

Philadelphia Aquarium Dolphin Show Tank

They had a 200,000-gallon tank with glass on the upper sides; it was where the porpoise shows were viewed daily. There was an electric overhead crane for lifting items into the tank, a perfect solution for moving our test subject in and out. Removing someone wearing a water filled space suit was a heavy job. I never knew whether we paid a fee or donation, but the aquarium was our next practice site.

When they closed for the day, our crew would help the aquarium staff relocate the bottlenose dolphins into holding basins at the rear of the building, and lower the gate. That was simple. Show a fish to a dolphin and they'll follow anywhere. At first, orientation tests were conducted without the space suit, and using a hookah type airhose. I had a removable, two-man platform built that was suspended from the tank rail. Once Fred was familiar with the test setup, he would suit up with everything except the helmet and pack. He'd move to the crane platform, and be lowered partway. Jack, or a technician, could stand on the side platform, finish installing the helmet, and pressurize the system. As the suit and pack performance became more reliable, Fred would leave the crane platform to do astronaut-like tasks while one of us monitored his body functions or movements. The test setups proved his ability to function normally. He was able to work while submerged for progressively longer periods of time, and we developed more difficult tasks for him.

The tank and holding pens circulated the same water, so the test subject and support divers were in the same water as the dolphins. That posed a problem only once when one of the dolphins barfed in the water. Fred wanted to proceed, but those who had to help him in and out of the tank said, "No way!"

Test Console Orientation
Fred Parker

Off-site testing was of help to me in establishing procedures for trips to Little Buck Island. All of the hand tools were fitted into tool chests. The rebreather and its supplies went into custom-built equipment trunks. I created checklists to be sure we had everything before we left King of Prussia. Dave Stith had presented each of us with a small o-ring when we completed his scuba training, the ring that was used to seal our regulator against the pillar valve on top of an air tank. It was his way of reminding us to check our equipment before we left home. The loss of something as small as a five-cent o-ring could cause waste an entire trip.

Two of Dave's Navy buddies stopped by one night; they were still in the military, as Seals. Hearing what we were doing, they wanted to see for themselves, and climbed into the tank with Fred for a close-up look. They wore contact lenses in lieu of facemasks. Oversize in diameter so they wouldn't slip out of the eye too easily, the striking feature was a fluorescent ring. It would help to locate the lens should it fall out in the water. The color was outside of the field of vision, and didn't interfere with a divers view, but the rings presented an eerie look to everyone except the wearer. One of the Seals gave Dave a pair of the oversized contacts for us to use instead of the goggles that were needed with our water-filled helmet.

Fred tried them later. "They aren't comfortable for me," he said. "My eyelids feel as though they are being stretched to hold them in place."

The Purchasing office delivered contractor bids, for manufacture of piece parts for my simulator, and I reviewed them. A firm in Westchester, Pennsylvania, gave the best price; their building was only twenty miles from the office. They had a massive rolling machine, well able to handle structural angles. I gasped the first time I saw the segments that were rolled; twisted out of plane, they looked like pretzels for some monster's snack.

I was incredulous, and asked the foreman, "Can you straighten them? We can't afford the extra time to remake them."

He laughed. "Rolling distorts most structural shapes, they're thick and thin in different areas. This is just the way they look before Kenny gets them."

"Kenny? Who is Kenny?"

"Well—he's out of the building right now, but he's an artist with a hammer and anvil. Used to be a body shop man, but we pay better, and he likes the challenge. Come back the end of the week and look at them."

I returned on Friday, and the foreman took me onto the work floor. Pointing to a man standing near an anvil, he said, "There's Kenny. Watch him."

The sight of this tall, lanky individual, shirtsleeves rolled above the elbow, standing with a ten-pound hammer ready to strike, reminded me Longfellow's poem. The words flashed across my mind, "Under the spreading chestnut tree, the village smithy stands—"

Kenny stared at the ring segment for a few seconds, sighted along the edge, and delivered a few sharp blows on the reverse side from the twist. To me it seemed the wrong side to hammer, but as he moved down the length of the angle, inches at a time, the piece uncurled to usable shape.

The foreman came up behind me again. "You see why I said Ken is an artist; look at the satisfaction on his face. Funny thing, I can't use him on a lot of machines; he's slow or clumsy on them, but not here."

Rocky's technicians completed the deck supports. Together with all the necessary nuts, bolts, and tools, the parts were flown to St. Thomas. Airlift International is a cargo airline that lands there, three days each week. I returned to the island with four of Rocky's technicians. On this trip we stayed at Bluebeard's Castle, a hilltop resort. Breakfast and dinner was included with our rooms, but while at the island we needed to have something to eat and drink. I arranged with the kitchen for box lunches to be packed and ready when we finished breakfast. The crew asked me to include soft drinks; by afternoon it was obvious that water would have been the better choice. That first evening we returned early and I bought water jugs.

We all wore tee shirts, swim suits, or shorts because the weather was so warm and sunny. On the first afternoon, a cold, sleet like, twenty-minute rain caught us flat-footed. With no shelter to protect us, the U.S. Navy sign fell victim to our need; it was pulled down temporarily, so we could duck underneath. Every day, cold rain came at about three o'clock, for twenty to thirty minutes. But after the first day we were better prepared for it.

Concrete cures hardest when exposed to water, and those twin strips were well cured. Broken or not, they resisted our activity. Large gravel had been used for the aggregate part of its mix whenever the strips had been poured, and we had to shatter some stones with a cold chisel before we could drill past them. We were miles away from any electric plug, and there were few cordless drills manufactured at the time. In any event, short battery life and the need for frequent recharging eliminated that option. The only way to drill those anchors was by hand, using star drills, ten-pound hammers and brute force.

Everyone had painful blisters before the job was finished. It remains a curious thought of mine, to remember that for the first steps of our great new technology we had to drill those anchors in the same way mankind has been making holes in stone for centuries. My part in the platform construction was finished. The next step was for the contractor to barge prefabricated sections of the building out to the site, erect the twelve by twenty-foot building, and install a pair of three-kilowatt generators.

We flew home late on Sunday night. At the terminal in Philadelphia I reminded the gang that we were back in the land of right hand driving. The first turn from the parking lot leads to Airport Circle, a well-known scene for accidents. I didn't want anyone to drive into traffic in the wrong direction. I slept the peace of the just, feeling that we had done a yeoman's job. Monday morning I ate a late breakfast with the family and wandered in at nine thirty. No one else was in my office, but that wasn't unusual. Like as not Charlie and Jack would be in Rocky's lab. A package of drawings for the simulator cradle was on my desk. Bob had left a note that these were my file copies; he'd completed them as I was leaving on the last trip. The cradle drawings hadn't been started when the original bids went out for the simulator.

Fred wandered in and sat down. His lower lip jutted out, as it did when he was pensive. He asked, "What do you still have open with the contractor in Westchester? We have to see Williams in Purchasing, this morning. I need to know where the job stands."

"They've given us a tentative date for delivery of the simulator parts. The only thing pending is this cradle." I picked up the drawings. "The vendor has a set of these drawings so he can make his quote; Bob lives so close to their plant that I asked him to deliver a set for me. It's a simple design, I'll call and get his price."

I stretched my arm toward the telephone, but Fred waved his hand, and said, "Don't bother. That's what I needed to know. Let's go see Williams."

While walking down the aisle Fred told me, "It may have been your intent for the vendor to look the drawings over and give a quote, but, he proceeded to make them. It was when they called for an extension to the purchasing order that all hell broke loose. In a large company everything requires paperwork. You've been accused of committing the cardinal sin of omission. Williams says that you violated purchasing procedures, and wants you fired."

I stopped in my tracks. Fred just called back, "Come on, times a wastin'. Let's see what mood he's in today." Catching up with him, we marched down the hall together.

Fred introduced me to Williams, a pot-bellied man who couldn't be bothered to shake my hand. His charges just started flying; Fred and I sat silent. "Contact with a vendor outside of channels—reckless disregard—performing favors—"

Fred held his hand up. He asked me the same questions as before: when were the drawings ready, why had I sent them as I did, what I'd said to my courier, and what the vendor was supposed to do. Then he summarized. "Drawings sent were to be reviewed, not built. The $1,200.00 for new work is so small compared to $58, 000.00 for the basic contract that there is no question of granting favors. You know that it's not out of the ordinary to add a small amount on an original work order."

Fred turned to me again, and asked, "Did you try to contact anyone in Purchasing before you sent the drawings out?"

"Yes, sir. On Friday, but I kept getting busy signals."

"Did you try on Saturday?"

"Only once, but the line was still busy."

He tented his fingers together, leaned forward toward my accuser and lowered his voice. "Do you remember what was going on in this office that week? I do." The pot bellied man looked puzzled and Fred went on. "Purchasing's phone lines were being re-laid under a false floor; the phones were tied up on and off for most of last week. Now, does that answer your damn fool questions so we can get back to work?"

Fred rose and started for the door. His face looked indignant. Williams began to speak, but stammered. It might have been an apology, or he may have been upset with his show having been concluded. When Fred went for the door I followed close on his heels. I wanted out of there, but to tell the truth I also felt smug at the way Williams had been put down.

At the first corner we turned. As soon as we were out of sight, Fred lifted his forearm, caught me under the chin, and pushed me against the wall. He stared at me eyeball to eyeball.

"Do you understand that I just saved your ass?"

I nodded. "Yes," I croaked.

"Well I was able to save it because we were dealing with an egotist. I found out about the phones by accident and they weren't off all of the time. Besides, you could have strolled down the hall and saved us some trouble." His eyes narrowed to a squint. "There's less progress when there's waves. Don't go around making waves. Kapisch?"

I nodded again.

Fred pulled his arm back. "Cover your ass in paper," he said. "We don't have the luxury of time to waste on this kind of crap." He strode back to the office with a humbled engineer in tow. For a small man, Fred cast an awfully big wake.

The simulator parts were ready for shipment to King of Prussia when we were informed that another program would occupy the workspace we had expected to use. One alternative was to work outside the building. Frost arrives in Pennsylvania anytime after Labor Day, and it would soon be nippy if we planned to work out of doors—very soon.

4

"Half of the world does not know how the other half lives."

—*Gargantua*, by Rabelais

Carl Cording was tall, thin, blonde, and handsome, but it was his smile that made lasting friendships. He was an old friend of the GE manager who ran NASA's rocket test site near Picayune, Mississippi, and had spoken with his friend about our dilemma. Now, that man wanted to hear a proposal about doing our subassembly work in one of their huge buildings. I had an hour's notice to collect the drawings needed and get to the airport. It had become my habit to keep a duffel bag in the trunk of my car. Carl had one vehicle laid up in the repair shop, and asked whether I'd drive so his wife could keep their other one. It wasn't a problem for me.

Rain had been falling hard earlier in the day but by two o'clock it had eased off. Carl was standing in the front lobby, suitcase in hand. He opened the rear passenger door and tossed his bag on the floor between the seats. The trip to the airport, the flight, and obtaining a rental car went all right, but we arrived in New Orleans without knowing if we had firm hotel reservations.

Carl didn't rough it when he traveled. He liked the Roosevelt Hotel in the French Quarter. It had a remnant of earlier glory; still plush, but with a few moths eating at the edges of the carpets and drapes. Upon arrival, we found that only one room was reserved. I'd bunked double before, but never with a manager. Usually they wanted their own space, but it would have to be so. Once sure of lodging, we walked to a late dinner in the French Quarter. Antoine's was a first class restaurant whose specialty was fish—light and flaky, but dry to my taste. Carl insisted that a wine would enhance it, and the side dishes, though heavily spiced, were tasty. I thought for a millisecond about the expense for our food and that particular hotel, and then promptly forgot about it. Carl was the one who had to approve my expense account.

Afterward we strolled among the taverns and music houses owned by people whose records I loved: Pete Fountain, and Al Hirt. By the time we arrived back at the hotel I was ready to sleep. When Carl opened his bag, everything in it was soaked. On the floor, in the rear of my car, his suitcase had landed in a puddle of water that had leaked in from somewhere—maybe a bad window seal.

I began to apologize, but he interrupted.

"Forget it," he said. "I can wear these same clothes tomorrow. Any time a day goes poorly, I figure the next day will be great." Carl had to be the eternal optimist. "It's two hours to Mississippi Test. Let's get out of here at six a.m. and eat on the road."

Interstate highway I-10 was not complete at the time, but sections of it were under construction. We drove on the old roads. Thirty minutes east of the city the blacktop narrowed to two lanes, and proceeded along the top of an earth dike that was six feet higher than the marshy fields around us. Carl noticed my curious glances at our surroundings. "Southern Louisiana and Mississippi are hardly above sea level," he said. "Those houses aren't on a regular foundation—they're built on pilings."

"There are rowboats sitting on dry ground," I replied in amusement. "Tied to the front porch. Ready and waiting."

The countryside became monotonous, consisting only of wet fields, huge trees, and Spanish moss. When the fields rose and were level with the road, Carl pointed to the front. "There's a café up ahead. We'll stop there."

My eyes scanned the horizon and found a small wooden structure; it appeared to be in the middle of the road. As we came closer, I saw that the road branched to both sides. It was a carbon copy of the greasy spoon café depicted on the Hee-Haw television show. Twenty cars in the parking lot made me think that we'd come on the day they were shooting next week's episode.

Carl led the way inside. There was a row of tables near the windows, but a horseshoe shaped counter surrounded the kitchen and serviced half of the chairs in the room. The menu spelled out plain and simple fare. I ordered one of the breakfast specials: steak with eggs, and coffee that I expected would be at least half chicory. As the middle-aged, plump waitress took our orders, I was struck by the resemblance to her TV counterpart, Lulu.

When my plate arrived, it also contained something I hadn't expected—a huge white mound. Not comprehending what it was, I stared at my plate. The waitress saw my puzzled expression. She leaned over, laid a ham-sized forearm on the counter, and stared at me until I looked up. "Somethin' wrong, son?"

I pointed to the mystery object. "Just wondering what that white stuff is."

She snorted. "Them's grits, son."

"But I didn't order grits."

"You don't order grits, Yankee boy," she declared in a loud voice. "You jes' gets grits." With that astute observation, she moved down the counter to intimidate someone else.

Carl laughed, but not loud. "If you don't want to wear those grits out of here, I suggest you doctor them up with butter or sugar and eat them. O God! I'll have to tell Fred what your face looked like just then."

One thing about working for a company the size and diversity of General Electric, is that you can move within the company to where there is more or better work. It's possible to walk into a building a thousand miles from home and meet people from the past. When I entered GE's drafting and engineering office at Mississippi Test, there were fifteen people I'd last seen at Naval Ordinance Department, Pittsfield. Some of them I had interviewed for Phil Weissbrod when they were hired.

It was a good meeting with the GE management team. Before leaving I called my vendor in Westchester, and arranged to have the parts delivered to Picayune. Carl and I returned to King of Prussia believing it was a done deal. Two days later the daydream burst, and I made a quick trip back to salvage whatever I could. On that occasion, I cut east across the city and drove over the Lake Pontchartrain Causeway. A narrow ribbon of road and railings twenty-six miles long, at least twenty of those miles are out of sight from land and skimming the lake surface. It was a different kind of driving experience.

The first thing that I had planned to do when the parts arrived at Mississippi Test was to dip-paint the structural pieces with two coats of algae resistant paint. That required construction of paint tanks, and racks on which the parts could be hung to dry between coats. The plumbers and carpenters unions chose to disagree on who would build the racks. We didn't care who won, but couldn't spare the time to debate it. GE base management didn't want labor problems; they decided to cancel what was, to them, our tiny project.

Thanks to a helpful purchasing agent at Mississippi Test a solution was found at W&A Engineering. Located near the airport in New Orleans, they had just completed a project and cleared a tall, spacious, indoor area. It was large enough to set up three of the structural ring sections at one time. The GE management group also loaned me a mechanical engineer to stay in New Orleans while the project was in construction at W&A. It made it easier for me to have someone at the contractors shop, and minimized the time I needed to be on that site.

W&A's management sat with me, to review the assembly drawings. They made a few recommendations to ease the assembly procedure. To keep the ring flanges from having mismatched boltholes, they suggested construction of a twenty-two foot diameter template. I agreed, and a twenty-four foot square plywood platform was built in four by eight-foot sections. Cast in small pad areas, West Indian's dock was by no means level; I'd reuse the platform there to assure flatness and curvature of each cylinder as it was reassembled.

The workers constructed long, narrow paint trays in which parts could be dipped, but no racks. The parts were hung from the rafters to dry, with plastic sheeting on the assembly floor to catch the drips. I spent one or two days in New Orleans each week, to oversee, and resolve problems. For the next seven weeks, it was a Sunday night flight or the Monday 5:30 a.m. redeye for me, arriving at 7:00 a.m. local time. I returned to King of Prussia as soon as possible, to work on other designs; there were more than enough other things that had to be designed and built.

Working on projects for a huge company, you get attention in a hurry by throwing the company's name out as an introduction. I picked up the phone to speak with a vendor, and said, "I'm Eric Bishop with General Electric Company and—" I didn't get any further before he interrupted with, "General Electric? Ain't you the folks that make those washing machines?" In that region GE was definitely not viewed with the same awe as a major local employer.

Once the first cylindrical section was built, and W&A's personnel were able to see what each segment and section looked like, their learning period was over. It was only necessary to build more of the same. I could be less long or less frequently in attendance, and communicate by phone. Fred and I met with NASA officials at Mississippi Test on three occasions when they were invited to review our work in New Orleans. I was privileged to meet Gordon Cooper, Gus Grissom, and Deke Slayton—three of the original seven astronauts. At two meetings in the W&A assembly area, I met four pilots selected for the Manned Orbiting Laboratory; Air Force didn't use the title astronaut. Because of security I was given only their first names, but the MOL team reviewed any projects that could benefit long-term space travel.

On four occasions I flew from New Orleans to St. Thomas to firm up contracts. Once I missed the last flight connection from Miami, Florida, to San Juan, Puerto Rico, and called my brother. Warren's family was moving there but the others hadn't arrived. He was staying with his wife's sister, Carolyn, and her husband, Manny; I'd attended elementary school with Carolyn. The guys picked me up from the airport, and took me to their home. Everyone listened intently as to what our underwater crew had done so far, and what was intended for the future. In the morning I continued my trip to St. Thomas.

Whenever I was back at King of Prussia, I worked on whatever part of the program was needed next. Some visitors to Little Buck Island were not expected to have experience as divers, or even to be swimmers. An observation tower was designed with its lower edge to sit below the water's surface. Visitors could view test and training, without getting wet. The tower scaffolding was to be held in place with sandbags on the ocean floor, and the corners cabled to anchors. Rocky put those parts out for bid.

The suit and helmet were water filled when pressurized, and to fit a scuba mouthpiece it had been necessary to remove padding around the mouth area. Team members held a meeting to see whether a communication system could also be installed in the helmet, but it was obvious that we were trying to stuff seven pounds of gear into a five-pound space. Jack Burt also wanted to go on record that he thought helmet removal was too slow and awkward if the test subject developed any kind of emergency.

Turning to Fred, he said, "I'm not sure how fast we could remove it and get air to you."

The upshot of that comment was for me to design a non-standard helmet that would mate to the space suit neck ring, but fit the project needs. I located a vendor in Connecticut who manufactured plastic spheres, and was asked to visit them. They had molds for various diameters, and worked with several different materials. My choice was Lexan, since it wouldn't scratch over time. To soften the material, plastic sheets were placed in an oven and heated for hours. A sheet was then clamped in place over a mold, and a vacuum line opened; the flexible plastic was sucked into a half-spherical cavity mold. As a first sample to try, they gave me two halves of a sixteen-inch sphere.

I designed a pair of aluminum rings to clamp the hemispheric sections together. The rings incorporated a small o-ring seal to prevent loss of internal pressure. Quick release was essential, and the connector chosen was a Marman clamp, a v-shaped band. It held the helmet rings together by means of an overthrow type of mechanism, and released by an easy tug on the handle. But where should the helmet split? It had to be above the neck ring, but low enough to allow access around the mouth or face. I tilted the sphere back and forth until I was satisfied. The bottom edge of its diameter would be one inch below Fred's chin; the top was at the back of the skull, for maximum facial access. When the first unit was complete, Bob Plank took my notes and sketches so he could document them. Then I left again for New Orleans.

Once W&A had built those ten ring sections, and the segments mated against the plywood template, all sections were numbered and dismantled for shipping. Marking of the parts was critical, to prevent having a large jigsaw puzzle at the other end. Open sided wooden crates were built to firmly hold the ring segments. Every piece was essentially a custom fit—damaged parts couldn't be repaired or replaced easily. At the end of September the cargo was loaded onto a freighter for shipment to West Indian Company at St. Thomas. I returned to King of Prussia and accepted a bid on the observation tower piece parts; they were to be built for later shipment. Now it was time to gather tools and hardware that would be needed to reassemble the simulator structure at West Indian's dock—starting in one week.

The shipment from Bay St. Louis, Mississippi, arrived at St. Thomas in twenty crates, and the wooden assembly platform was in eighteen four by eight-foot sections. My cargo consisted of more bulk than weight. Sunday afternoon I arrived with Bob Plank, Eddie Sienko, and Frank Pugliese who would be taking photographs while we assembled the structure at dockside. Later, he'd do the underwater filming. Ted James, owner of a car rental located on the ground floor under Sebastion's Cafe, met us at the airport to deliver two Volkswagens that I'd rented. My cargo was where I expected, behind the watchman's shack at West Indian. I had a packing invoice, and checked that all of the W&A shipment had arrived. So had the hardware and tools I'd sent from King of Prussia. Everything had been stored unopened and I left it that way for now.

In 1917, the United States bought part of the Danish West Indies. Later known as the U.S. Virgin Islands, the sale included St. Thomas, St. John and St. Croix. Denmark received fifty million dollars, but it prevented the islands from being used to rearm and refuel German submarines that had been wreaking havoc with shipping in the Atlantic and Caribbean. Off-season prices change in mid-December when tourists, like the birds, flock South; prices revert back to half-price one week after Easter. There are more than sixty recognized holidays each year: Christian feasts, saint's namedays, a week of carnival after Easter, voodoo celebrations, beginning of hurricane season and the end of it—when an employee can request time off. They don't have to be paid for all of those days off, but if they do work it may require overtime pay. I had to approve several of those holidays as overtime for the West Indian Company workers, just to maintain the schedule.

Charlotte Amalie is the only town on St. Thomas, and is named for a former Danish empress, Charlotte Amelia. The road around the perimeter has a speed limit of thirty-five miles per hour, the highest allowed on the island. Gasoline is more expensive than on the mainland, and the majority of cars are economical models—primarily Volkswagens. Driving is on the left-hand side of the road, but in vehicles with U.S. style steering wheels. Nonetheless, it wasn't hard to get used to the difference.

A ridge of heavily wooded mountainside separates the north shore from town. Narrow roads cut across the slopes like a roller coaster. Twists, turns, drops and climbs are so frequent that a driver can hardly get out of second gear. Morningstar Beach Motel is located at the southeast corner of St. Thomas. That's where I'd reserved rooms on this trip—for an indefinite stay. It would be cheaper to work from Morningstar than Bluebeard's Castle. The motel had a bar, but no restaurant; we could eat where or when we pleased, rather than accommodate resort community dining hours. Today the long single-story building that once fronted the parking lot, with doors that opened onto the white sandy beach, has been rebuilt into multi-story buildings. Only the beach and sea view is the same.

A narrow ribbon of blacktop winds over and around the mountain that sits between Morningstar Beach and West Indian Company. If the mountain has a name, I never heard it. A new hotel, Frenchman's Reef, sits on top of the mountain today, but the location provides the same panorama view with breath-taking scenery.

West Indian is an import-export shipping company, another hold-over from the days when Denmark ruled St. Thomas. Their pier is the docking site of freighters as well as passenger ships, and it's long enough to handle three small ships at one time. The newer, larger, cruise chips anchor in the mouth of the bay and transfer passengers to the dock by motor launches. The morning after our arrival, we had breakfast in town and arrived on West Indian's dock before the sun was up. The mountain on the east side of the dock prevents any gradual sunrise. We watched in awe as the first inkling of light broke the dark; within seconds the sun topped the mountain. It was like pulling a lamp cord.

Workers in St. Thomas aren't always locals. The lower Antilles chain of islands are very poor, and their people come to the larger tourist islands for better paying jobs. Migrants have seventy-two hours to locate a job or go home. If fired, they have forty-eight hours to locate another position or, again, they leave. Down-islanders stay for long stretches of time, to go home would use precious dollars. The majority of them send money to families they may only see once or twice a year. With the threat of deportation hanging over their heads like the sword of Damocles, they make subservient workers. Used by an unscrupulous employer, it can be another form of slavery.

To aid in uncrating and assembly I'd contracted for two West Indian machine shop workers: Daryl and Theo. Daryl located a forklift truck, and positioned the crates for me. Our wooden assembly template was rebuilt, and shimmed to level it. The most noticeable thing about Daryl was size—both his height and weight. A huge man, what might at first seem to be fat rippled with musculature when you looked closer. A gentle giant, he approached me with his eyes cast toward the ground. He said, "Sir. What is to be done with the wood from the crates?"

It wasn't something I'd thought about. The wood could be discarded. "Will the company haul it away? Or should we burn it?"

"Oh! No, sir." Daryl was horrified to think that it might be burned. It took a moment before he spoke again. He pointed to his coworker. "Theo is my wife's brother, and lives with us. He wants to marry. It would build another room, if I could have some of it."

"You're welcome to all of it," I said. "We won't open all the crates at once. That will give you time to move it in small loads."

His grin would have lit a ballroom. "Yes, sir. Thank you very much, sir!" He backed away, and walked rapidly toward Theo, apparently to break his news.

The two men placed every board within sight of where they worked. They had brought their lunch. While we went to town, Daryl and Theo removed nails, and stacked boards. They even collected bent nails, to be straightened later. I informed the watchman that the two workmen had permission to take the material home; he seemed disappointed that the lot hadn't fallen to him.

After work, Daryl made three trips in his old and weary Volkswagen. Boards were tied on the roof in a precarious manner, and others stuck out from the windows. I don't know how he managed to drive from the position that left him, but where there's a will there's definitely a way. I staged the uncrating in intervals that allowed them time to move all of the crate material. Later, they borrowed a truck and removed the platform sections.

Dockside, West Indian Co.,
St.Thomas, USVI
Author, Right Foreground

Using two saw horses and a sheet of plywood; I made a field desk for myself and unrolled a set of prints. Eight ring segments were soon laid out on the wooden platform, and bolted together as a ring; each cylindrical section weighed about six hundred pounds. My crane contractor wouldn't be available until the next week. We had brought furniture dollies and slid each finished ring off the platform, and onto the dollies. That freed the platform for another ring to be started. Six were completed before the crane arrived.

The fifty-five foot boom length was the longest available on the island, and had two lifting hooks; that would make rotating the ring cylinders a simple 90-degree turn to an upright position. When the assembly was complete, the simulator was to be lifted from the dockside onto the barge. Next, the crane would move to a dirt ramp and maneuver onto the barge. The crane was required again at Little Buck Island, to offload the structure into the water.

My crane man presented a problem from the beginning. When I'd first seen his rig the two hooks worked. But when he and his crew arrived on this occasion, only the top hook was in working order—the other had a cable tangled in its pulley. Both hooks were needed if he was to turn each ring. I told him that we expected him to perform to contract—to get the other hook working. He didn't want to take the time for maintenance; if I insisted he'd just move on to his next job. I needed him more than he needed my work.

Preparing for a Crane Lift
Top: Left, Bob Plank
Right, Ed Sienko
Bottom: 2nd Left, Fred Parker;
Center, Author
With West Indian Employees

Cable movement over the working pulley wasn't smooth; the line jerked whenever the crane started and stopped. My cylindrical ring structures were as delicate as eggshells, and the operator behaved like his crane was a giant yo-yo. The first cylinder bounced so severely that I felt the bolts would snap. I ordered him to put it down until the joints could be reinforced.

The quickest thing I could think of, to absorb alternating tension and compression loads that pulsed through the cylinder as the crane jerked and pulled, was to tie nylon cords diagonally across the ring sections. Lots of cords. It was also apparent to me that a single ring could not be turned upright; it needed to be bolted to a second unit so the joints were reinforced. I had Bob and Eddie lash boards across a second cylinder, to brace it for a horizontal lift. Once the operator raised it five feet, it was placed over the first ring section. When the two were bolted together, and the eight ring joints were overlapped, stiffness of the sub-assembly was improved.

Instead of picking up the paired units at their diameter to pivot them upright, now it was necessary to attach cables to the topmost point. They were lifted and placed on the cradle. Not knowing when it would be safe to cut the diagonal cords, I left them in place. They wouldn't be in our way, and it lessened the potential for breakage.

Despite all the precautions I could take, rough crane handling did crack eight joints. When the first fracture occurred, I didn't dare allow the crane to either raise or lower the ring sections. Splice plates were made in West Indian's machine shop, but to attach them I needed a long ladder. The only ladders around the dock area were homemade, pieced together with scrap boards. None of the crew would be safe climbing up twelve to sixteen feet on that kind of junk. I took Eddie Sienko with me to a hardware store not far from the West Indian gate, and purchased the best one that could be found on the spur of the minute. It was a twenty four-foot extension ladder that cost sixty-two dollars. Weeks later, when I returned to King of Prussia, I filed my expense account.

The money taken with me was for personal expenses, but it had been necessary to spend it on hardware or the project would stand still. I felt that it would be a simple matter to transfer the cost of the ladder, to the equipment account number I provided. Hah! GE's accounting office had question after question about why I'd buy hardware with money intended for living expenses. Fred called them for me, and tried to straighten it out. More questions. They refused to transfer the item. Eight months had passed and I began to wonder if they would take that sixty-two dollars out of my first retirement check. Carl called them. As a personal favor to him, the account manager finally agreed to transfer the item from my personal expense record.

But once more, when my copy came back, there was another question. The accountant now wanted to know, "Did you go out for competitive bids on the ladder?" How do you answer someone with tunnel vision? Two hundred and fifty dollars a day for the crane, plus salaries of the West Indian and GE workers waiting for me to get back with a ladder. How important was it that sixty-two dollars was spent out of the wrong cash pocket? Only a bean counter, someone that never leaves the office to see how the real world operates, could possibly have an answer for that.

On the West Indian work site, bright yellow tapes were strung around the work area. They warned debarking passengers not to walk there. Many people came off the liners and gawked at the big yellow cage. One or two of us would stop other activities, to keep them from blundering through our work area while on their way to the front gate. We fielded uncounted questions about what the structure was, but our main concern was to watch for the occasional dingbat who tried to walk past us on the outside edge of the pier. A misstep on that side would result in a twelve-foot "Steve Brodie." I was sure that a fall into that murky water wouldn't kill anyone, but we didn't need lawsuits.

From the moment a liner arrives, until the last purchase is delivered or passenger returned, activity among the local entrepreneurs continues at a frenzied pace. Tourism is the mother's milk of St. Thomas. In choosing this location we were exposed to the exotic nature of the Caribbean, and doing business where half of the population take nothing serious. One good example was a failure to receive a cargo from Airlift International. As happened too often, their plane didn't come in on schedule, or the next day. Their office couldn't or wouldn't, tell us whether the cargo had even arrived in San Juan, Puerto Rico.

The clerk at the St Thomas office just said, "No big t'ing, mon. Try again tomorrow or nex' day." Days of that, with the clock and expenses running, became frustrating.

GE's Transportation Manager in King of Prussia, Paul Taylor, solved that with surprising ease, and later boasted to me about his conversation. He placed a call to the president of Airlift, a gentleman by the name of Block. General Electric being a large client, he took the call.

Paul asked, "How do you spell your name, sir?"

Astonished at the question on such a short, easily spelled name, Mr. Block poured forth the letters, "B-l-o-c-k. Why?"

"Well," answered Paul, "when I send my report to CAA, I want to make sure I've spelled your name right."

Spluttered questions on the other end indicated success; Paul had Mr. Block's undivided attention. He told how difficult it was for us to work in St. Thomas when their plane schedule failed. Airlift didn't want large, unhappy customers, or federal regulators looking at schedule failure. My cargo arrived before ten o'clock the next day.

Lars Pederson was a Danish citizen. Not a physically large person, his strength was in his personality. He had a leathery face and a piercing look to his eyes, but it was his voice and words that had the effect of cracking a whip. One afternoon as we left his office, he reached for a large-brimmed straw hat and said, "I've had too many skin cancers to go out without this." He smiled, and asked my opinion. "Do you like my hat?"

Its broad brim had a feminine look, but I didn't want to say that. "If it works for you it's okay. Has someone been kidding you about it?"

The smile vanished, and his eyes turned into slits. He almost hissed when he responded, "No one on these docks ever laughs at me." We passed a crate sitting alongside one of the sheds. As if to demonstrate his earlier remark, he called one of the workmen over, pointed to a building and said, "I told you that package was to be put inside the shed. If I see it on my way back, I don't want to see you."

That worker called another man over, and together they struggled to make the heavy, offensive crate vanish. Pederson was always polite to me, and we had a good work relationship, but I never felt comfortable enough with him to call him by his first name.

One local person helpful to our task was the U.S. Commissioner of Water Sports and Recreation; I believe his name was George Morgan. What visions his title brings forth. In reality he was a hard working GS-11 federal employee, whose job was to organize diving, swimming, fishing contests; in general, he created and publicized events to go with tourism. He assisted us in many ways, the least of which was to find Al White. At first, when we were without an office, he allowed us to use his until we rented one.

John Bennett was one of Al's part-time divers. He had converted his hobby of diving into a career of diving lessons at Magen's Bay hotel, on the north shore. John was a former Marine—if there truly is anything former about any Marine. Those I've met still carry so much of the military in their bearing that it's like they never left the service. He still bore the buzz haircut, and there was formality in his voice. Good man, a great diver, but brusque.

An exact opposite type was Basil, a professional diver for hardhat and scuba. More like Dave Stith, he was fun loving, easygoing, everyone's friend. Basil and Al had worked together recently, checking AT&T's underwater cables for insulation damage.

Al told us, "On one occasion Basil surfaced and yelled for help. He went back under and we thought he was in some kind of trouble. All of the other divers responded, and found Basil wrestling with two huge lobsters. He wasn't in any danger—just too stubborn to let one of them go."

Basil shared Dave's disdain for water as a consumable item. He held maritime credentials as first mate on English or American registered ships. One evening he told those of us at his table, "At the end of WW-II the British Admiralty wanted to determine the qualities of military divers. They set up a special camp to conduct physical and psychological tests.

"Everyone was kept on the base to control exercise, food, drink, even sex. So many of us went under or over the fence for a night in town that testing was discontinued. Their final report said, 'The best candidates for diving are not liable to put up with severe restrictions on their personal lives.' I agreed with them; a professional diver has to be a little bit crazy."

Many of the residents of St. Thomas also showed curiosity about the yellow cage being assembled at West Indian. It was such a large object, and brightly colored, that it could be seen from a mile out in the bay. One would have to be legally blind to overlook it. Carl Cording scheduled a news conference for all the media. One enterprising reporter from a local tabloid came the afternoon before the conference.

I saw him duck under the tape, and said, "Please go around, for your own safety."

He continued toward the structure, and I moved to intercept him. Stopping in front of me, he told me his name and what paper he was with. He waved in the general direction of the structure and said, "I'm here to get a story on what you're doing."

"There's a general press conference scheduled tomorrow," I told him. "Ten a.m. at the Commissioner's office. Both of my manager's, Carl Cording and Fred Parker, will be there to explain it all."

"I know that. We all got the same invitation, but I want to get a jump on the story."

"You may as well leave then. I can't tell you anything more."

Next morning his paper had a front-page picture of the structure, with a headline such as most papers use to announce a war. "GE plans Secret Missile Center Here." His story began, "Your reporter was thrown off of the West Indian Company property by a General Electric Company manager named—"

Airlock Tunnel Section, Yacht Haven in Background

He'd staged our run-in so he could have a lead for his story, but it was wrong on all counts: I wasn't a manager, and what we were building was hardly secret. From the dock next door, at Yacht Haven, he'd taken a photo of our structure through a sea of at least one hundred boat masts. It reeked of tabloid, stretch-the-truth yellow journalism—so much for a secret missile center. All of the other papers and radio media reported factually.

Before the last ring sections were fitted to the cradle, I looked into the timing and logistics of moving my assembled structure to the cove at Little Buck Island. We had to have an earth ramp built on the sandy beach near the airport to allow the crane to be moved onto a barge. However, the dry season had arrived. St. Thomas has one freshwater well. Located on West Indian Company property, it is drilled very deep, but has only enough flow to supply company requirements. Most buildings on the island have rain gutters, and run the water into cisterns. A desalinization plant meets part of the island's demand, and water barge runs to Roosevelt Roads in Puerto Rico provide the rest.

The barge I'd contracted for was now making twenty-eight hour runs bringing water to the thirsty people of St. Thomas. We had an open dated contract, but stopping their schedule would be like turning off their money machine. The barge operators wanted me to pay double our agreed price for them to honor the contract. Also, GE management had been adamant that insurance be provided for the barge on the way to Little Buck. The company had no desire to become a third party to liability resulting from, say, the barge running down a small boat while on hire to us. They'd had me sit with a corporate attorney in King of Prussia for a morning while Jones' Maritime Law was explained to me. Where ships or seamen are involved there are no limits to the damages that can be requested in court.

Jones' Law was one of the reasons we had a divemaster at the work site. If anyone was performing their normal duties, even on dockside we were covered by Worker's Compensation: Insurance in our home state, and Al White's employees through the policy he provided. Before we donned diving gear we had to inform the divemaster. Once logged in with him, we were seamen according to Maritime Law; if we broke a leg climbing the staircase to the upper deck, a lawsuit could be filed for an unlimited amount. It's obvious why the company needed to know what category each worker was in, at all times.

Another thing that the company wanted to prevent was third party liability: for example, say the barge were to run over a small boat while employed by us, GE could be sued because they chartered the trip. Coverage for our five-mile barge trip to Buck Island had to be obtained from a maritime insurer. I'd tried to set times and dates with all three: the barge, crane, and Lloyds of London. Each time plans went awry. Thanksgiving was approaching, and I hated to leave St. Thomas with my schedule falling apart. So near, and yet so far. We were all frustrated. But working twelve hours a day, seven days a week, the crew needed time off. My mood was glum, and I didn't want to quit trying. But I joined the others for a holiday trip home. Some holiday.

5

"Give me your tired, your poor, your huddled masses yearning to breath free...."

— *The Lady of the Harbor,* by Emma Lazarus

I had spent two frustrating weeks trying to get it moved, but on Thanksgiving Day 1966 the SIV-B simulator hadn't made it off West Indian's dock. When I returned to St. Thomas, how could I be sure that I'd have any more success than before? "As a kid in Brooklyn," I thought, "all I wanted to do was teach. How did I ever get into this?"

Turkey dinner was to be at my parent's new home this year. The house where I grew up had become too large to care for; they'd moved to Kimball Street, near Flatbush Avenue. I knew the Marine Park district from times we'd gone to watch barnstormer airplanes at Floyd Bennett Field, which later became a Naval Air Station. We had a two-hour drive from Pennsylvania; that provided me with plenty of time to reflect on the direction my life had taken.

Affectionately called The Big Apple, New York City is America's greatest example of the term Melting Pot. Both branches of my family came together there. Maternal great-great-grandfather, Mariano Balmas, arrived from Castile, Spain; at seven he'd never seen snow. The boy thought the flakes were sugar, and ran around the deck trying to catch them on his tongue. His family settled in Brooklyn. Later he married Mariana Patti, a cousin of the opera singer Adelina Patti who was immortalized in song as 'Sweet Adeline.'

Their daughter, my great-grandmother Barbara Antonia, was born in Brooklyn in 1855, first of the four Balmas children. Her married name became Hantch, they had two children, and she became a young widow; grandmother Barbara Antoinette Hantch married a grocery store manager, Howard Sheldon. My mother, Emma Reeve Sheldon, was the younger of two children.

Great-grandfather George Bishop arrived from Kent, England, a younger son in a country where by tradition the eldest inherits. In 1859 he married Jane Eliza Wray, traveled to the U.S. and set up residence in what is now the NYC borough of Richmond, better known as Staten Island. Of eight children, five survived their sixth year.

Grandfather Thomas Bishop married a Staten Island farmgirl, Minnie Hurley. Her parents had arrived a generation earlier from County Cork, Ireland. In his seventies 'old man Hurley,' as my father always titled him, remarried to a younger woman whom his children disliked. Pat and Mike left family businesses they'd been operating for years, and disappeared. I think everyone has heard a Pat and Mike joke, or story. It was grandma's letters searching for her brothers that inspired me to look into our genealogy.

Staten Island, near Hurley Farm
Rear: Uncle Eugene Hurley
Driver: Milton Bishop
Passenger: Eric George Bishop

With grandpa Bishop's engraving business on Maiden Lane, four blocks north of Wall Street in Manhattan's banking district, trips to be with his family on Staten Island were limited to weekends and occasionally midweek. No doubt that contributed to their small family size. My father, Eric George Bishop, was the youngest of three.

The Dutch settled Brooklyn in 1636, spelled at the time as Breukelen. Incorporated into a village in 1816, it became a city in 1834, and remained an autonomous city until 1898 when it joined NYC as one of its five boroughs in "The Great Consolidation." Downtown Brooklyn is an architectural wonderland of Greek, Gothic, and Romanesque Revival styles together with Art Deco and modern buildings. Brooklyn Bridge opened in 1883, and created an easy commute to Manhattan.

To be closer to his business in lower Manhattan, grandpa Thomas Bishop moved his family to the north end of Brooklyn in 1915. They settled in the Bedford-Stuyvesant district where brownstone mansions built in the 1820's were becoming apartments. At the time, access to a cold-water flat in one of those buildings required a letter of reference from a previous landlord.

Gateway to the NYC Financial District
Brooklyn Bridge

Fate placed dad's family on the second floor of the building where mother's family occupied the fourth. At nineteen, he was ten years older, but as Maurice Chevalier said and sang for years—time has a way of changing little girls. Years later, when they began to date, each family expressed petty reasons for disapproving of the other. Maternal great-grandmother, Barbara Antonia (Balmas) Hantch, cast a shadow over their future. She claimed to have researched the Bishop family tree and didn't like what she'd found.

"In France," she told her family, "their surname was Levesque, meaning the Vicar. It's evident that after 1066 they promoted themselves in church hierarchy to the rank of Bishop. Besides," she continued, "William the Conqueror recruited the dregs of the jails for his army." As well as her being Spanish, the paternal side of the family resented her air of superiority, which Barbara Antonia wore as obviously as a mantilla.

Nevertheless, mother and dad continued to date. In August of 1926 the lovers took a train to Elkton, Maryland, and were secretly married. Decades later, when telling the story to me, dad said, "When the train passed through a tunnel, smoke from the steam engine squeezed into the coach. I kept her wrapped in newspaper to prevent the ashes from ruining her rose colored dress. On the way home, we stopped in Philadelphia to pick up souvenir postcards from the U.S. Sesquicentennial Exhibit. That was where we were supposed to be all day."

In six months, they had another wedding in New York City; it was the only one both families knew about until decades had passed. The newlyweds also settled in the Bedford-Stuyvesant district. Ten years after mother and dad's elopement, there was a newspaper expose of Elkton as a marriage-mill. Many of those performing ceremonies either had no license, or failed to renew them. Mother and dad had accidentally chosen one of the few who performed legal ceremonies.

Elkton, Maryland Newspaper Headline

Through the Balmas branch of the family my brother is reputed to be a Marquis, Spain is another country where the eldest inherits. I've not found proof of the lineage, and at the current value of Spanish titles neither he nor I have thought much about it. But Barbara Antonia behaved in a superior manner all her life. An easy way to earn her displeasure was to cast some aspersion on her small claim to fame.

Brooklyn covers seventy square miles. Today, it is home to ninety-three ethnic groups, and one hundred-fifty nationalities, in eighty neighborhoods. When I lived there it was more like a small town; schools, stores, and churches were within a few blocks. The city was criss-crossed with train, trolley, or bus routes for transportation. Automobiles weren't common, and roller skates, bicycle, or the shanks mare were a common mode of transportation for children. Borden's milk wagon and several fruit or vegetable vendors traveled the neighborhood in horse drawn wagons. I remember those equine years with embarrassment—mother made me collect their droppings for her flower garden.

The house I think of as my childhood home was on East 16th Street: two miles from Sheepshead Bay for fishing, three miles to Brighton Beach where we swam in the ocean surf, and four miles to Coney Island. An island in name only, it was billed as the "World's Largest Playground." Grandpa Sheldon loved to take his Sunday afternoon walk on Surf Avenue or the boardwalk. Whenever I was asked to join him, I knew we'd stop for hotdogs at Nathans—the self proclaimed Hotdog King—or at Feldman's if gramps wanted a beer with his hotdog. If he had a little extra change in his pocket, we'd venture into one of the sideshows, the freak shows, or Vera Zorina and her snakes (he liked the ladies), or we often rode the famous sixty mile an hour, wooden roller coaster—the Cyclone.

1953 East 16th Street, Brooklyn
My Boyhood Home

Our two-family house was seventy-four years old, had five rooms on two floors and three small rooms in the attic; we moved there when I was four. The building had been a residence for the Chief Engineer and his foreman at a water pumping station, located one-half block away. When the station closed, the house was moved south to the middle of the next block, and placed on a 40 by 100 foot lot; that was a large piece of land compared to the a 25 by 100 foot size for duplex style homes. When friends hear of my birthplace they tease about scenes from the book, *A Tree Grows in Brooklyn*, which portrays life in tenement houses of the type we left behind in Bedford-Stuyvesant. East 16th Street had houses only on the east side; facing those homes, on the west side, was a grove of tranquility—like a private park, or playground.

Brooklyn-Manhattan Transit system, the BMT line, runs from north to south. Passing under the business district trains climb out of the tunnels at Prospect Park, near Ebbett's Field. One time home of the Brooklyn Dodgers, and now covered with apartment houses, the team's name was derived because fans dodged the trolleys in order to reach the ticket booths. Beyond Prospect Park station, train tracks continue a short distance in an open trench. Then the tracks rise above ground on a dirt embankment and bridges provide passage over the streets.

The slope we faced was lush with foliage: maples, pines, elms, white birch, and flowering shrubs. It was completely countrified, right down to the poison ivy. A four-foot high picket fence at street level wasn't a deterrent to the children, but at the top of the slope an eight-foot chain link fence prevented access to the train tracks with their deadly third rail. A tire swing and the platform for a tree house were already in place when I moved there. Like the Garden of Eden to me, that 16th Street paradise inspired my love of nature.

Everyone could have used more during the depression, but we made do. Children wore hand-me-down clothes. Families and neighbors passed things on, and there was no social stigma attached. I wore corduroy knickers with knee high socks until the eighth grade. My first pair of long pants, dark brown tweed, had originally been purchased at John Wanamaker's. My cousin and brother had both worn them, but I was ecstatic when they became mine.

Our neighborhood had sixteen children within the immediate block, with no organized baseball, football and basketball leagues as there are today. Official game rules were modified to accommodate the number present, ages, or physical size and ability. Everyone played—boy or girl. We played hockey with or without skates. At times there were three teams for stickball: one in the infield, one covered the outfield, and one at bat. To suit uneven size teams, or big and little kids, one side might have only two strikes and you're out, the other may need four outs per inning instead of three. Whatever worked. It was a lesson that prepared my generation to making adjustments in everyday life. Ours was a one-way street, but when automobiles became more common we added a position called "lookout." It was an appointed position. Sometimes it was for an infraction committed, but to be fair that position was also rotated between all of us.

The most famous of dad's work assignments as a bridge painter is Brooklyn Bridge, but there are twenty-two more that connect the boroughs. Circle line cruises originally started from the Battery; today they leave the 42nd Street-Hudson River pier. After a view of the Statue of Liberty and Ellis Island near the southern tip, the boat turns north into the East River to travel the length of the Island. Then it squeezes through a narrow passage at Marble Hill Straits. Back on the Hudson, it travels south to the pier.

A Tour Boat Passing Brooklyn Bridge

At the end of the Thirties decade, about the time that the depression began to loosen its hold, a World's Fair was held in New York City. Newspapers ran stories about the progress, showing pictures of the exhibit buildings rising from the soil of Flushing Meadows. Daily stories told of the countries and corporations who would sponsor exhibits.

A trilon and perisphere became the symbol for the fair, and the Sunday paper always carried a report on the latest additions to an already long list of exhibitors.

Our family planned each visit to the fair more carefully than Stanley did when he searched the Congo for Dr. Livingston. A map of the grounds was consulted. First choices of the exhibits were decided in advance, plus an alternate or two in case of large crowds. Cash money was limited, but two amusement rides were allowed—one selected by each child. On the great day, we helped mother to wrap sandwiches in wax paper. Tubes of salt and pepper were rolled. Tomatoes, hard-boiled eggs, and fruit, everything was fitted into shoeboxes. Then dad would tie them closed, and weave a cord handle so the boxes could be carried comfortably. Two boxes were carried at the start, and repacked into one when lunch was over. That way we could stay past dinner.

One of the greatest exhibits, one that we saw on every trip made to the fair, was General Motors' "I have Seen the Future," where a conveyor carried seated viewers past models of cars, buildings, and clothing anticipated in later 20th century. Warren and I collected buttons from every exhibit that offered them; my favorite was the Heinz Company's pickle pin—I owned a dozen of them. We wandered pavilions until dark, watched fireworks, and left as the gates closed. It was a wonderland, an adventure for the entire family.

When I was eleven, dad took me for a close-up look at Brooklyn Bridge. Starting from the plaza in front of Borough Hall, we walked across the pedestrian path to Manhattan and back. He told me, "Its tall enough below the bridge for aircraft carriers to pass under as they make their way into the Brooklyn Navy Yard, and two hundred and seventy-four feet from ground level to the top of the towers." Each place where we stopped, he'd describe how or where his crew would hang scaffolds and bosun chairs. "But the surface has to be prepared first," he said. "Old, flaky paint has to be removed before we pick up a brush. Prepare and prime it, then it gets a full coat."

"How long does it take to paint it?" I asked.

"Two years for one coat, with the crew size we have now."

In September 1941, I overestimated my climbing ability and fell from a tree. The following day I wore a cast and arm sling. I attended a Catholic parochial school, and the sister-teacher had assigned a composition due on the next day. She made a great show of counting those papers turned in.

"I'm one short," she declared in mock surprise. "Whose work isn't here?" But she was looking straight at me.

I freely admitted to being the culprit. For emphasis I waved the sling in her general direction. "It's my right wrist. I can't write for awhile."

"God gave you two hands," she responded. "Use the other one. It may not look as neat, but I expect that you'll turn in the assignment—tomorrow."

I was livid, and fumed about it to mother. It was to no avail. Mom agreed. Needless to say, I turned in a paper the next day. It was purposefully sloppy, but sister graded my literary effort, not the style. Left-hand penmanship improved slowly. Because of the incident I do most things with either arm or hand whether I'm painting, or driving nails. It was an early lesson in how to handle adversity.

In the summer of 1942 I experienced a first hand opportunity to explore New York City's sights. Dad had drawn visual images of exhibits at the Museum of Natural History. It's in Manhattan near the northwest end of Central Park. I told my parents that I'd like to see those wonders for myself. They didn't mind, but they didn't want me to travel alone. I convinced a neighbor, Phil Haydeck, to join me. He was only months younger than I was, and attended a public elementary school but in the same grade. We each brought a sack lunch, two nickels for the subway trip out and back, and another nickel for something to drink with lunch.

Mother wrapped extra pocket change in the corner of a handkerchief, and gave it to me. She always used the expression "in case."

"You should carry this in your handkerchief, in case you should develop a hole in that pocket." I don't know how often I've heard the old line about changing your underwear, "In case you get hit by a car and go to the hospital." At the time I wondered if she meant that it was all right if you got hurt, but not if you were shamed.

Mother thought and talked that way. She had the "Toity-toid" street Brooklyn accent, and called what went into the furnace "earl" instead of "oil," or said that you wear "poils" which everyone else calls "pearls." People thought it was quaint. A couple of her friends referred to her as "Little Miss New York." The city's name was emphasized as "New Yawk." I've lost my accent, but Warren still has it. So do both of his sons: John and Peter.

In the ignorance of youth, Phil and I expected to travel to the Museum one time and see everything. Hah! The rooms awed us: the height, the huge area they covered, size of exhibits such as the dinosaurs, and the sheer number of showcases with cards to be read. Phil and I decided to come every weekend and do one room each trip. There would be plenty to talk about in the fall when our teachers asked their annual question, "What did you do this summer?"

The Museum closed at 3:00 p.m. on Saturdays, and we finished off each day by walking to anything else worth seeing—Central Park zoo, the eighteen-acre lake, and an Egyptian obelisk called Cleopatra's Needle. Braver each week, we continued further south: Rockefeller Center, with its Radio City Music Hall; Times Square and 42nd Street, the theater district. Finally we reached as far as Macy's, Gimbel's, and the Empire State Building at 34th Street. Fascinated by this new world that we had discovered, Phil and I continued our weekend excursions. Within a year of Saturdays we had covered the entire Museum of Natural History, the Planetarium, Metropolitan Museum of Art, the Brooklyn Art Museum, and the N.Y. City Aquarium.

1943 was a major turning point. In spring, my maternal grandmother Barbara Sheldon suffered a stroke and was paralyzed on her left side. She needed help with daily living, yet it was necessary for gramps to work. Mother, two cousins, and two kind neighbors shared the effort of seeing to her care, and preparation of meals. Then, in June, dad fell fifteen feet from an unsecured ladder. The fall left him with a broken neck and a severe head injury. At age forty-seven he was alive, but permanently crippled. On Workers Compensation, the family paycheck was cut in half—to two hundred dollars a month. A merger of two households and paychecks solved both the Bishop financial and Sheldon healthcare problems.

Our family relocated from the first floor apartment to the second, so my grandfather and brother could use two of the attic rooms as bedrooms. The household did experience strain because there were some hard feelings between two strong personalities—dad and grandma Sheldon. However, they understood the need, and avoided open conflict. With an old house there were always repairs to be made, and at age thirteen it fell on me to become the handyman of the house. I've never been sure why it didn't become Warren's role; he was good with tools at school, but he escaped those chores. Dad would show me, or describe what was needed, and how to do it. Then I'd do the physical part.

Dad and I accomplished many things together; what he taught made me self-confidant. I'd tackle any kind of project: painting, carpentry, and sometimes create special tools that either were not available, or expensive. He had a workshop in the basement. It wasn't equipped with major power tools like a table saw or drill press, but I put all one of his hand tools to use and learned how to sharpen and care for them. Dad had special instructions on how to clean his expensive camels-hair paintbrushes. When I painted the rough stucco part of the house, he would only let me use his nearly worn out brushes.

It fell on Warren and myself to finance our own miscellaneous expenses: movies, other entertainment, and school supplies. With the war in progress, it wasn't hard for either of us to find kid jobs. At first I wasn't old enough to apply for working papers, and hung out around stores where tips could be earned. Delicatessens, grocery stores, and flower shops always needed delivery boys. And mom-and-pop storeowners were happy to find someone who would restock shelves, but wasn't on their regular payroll.

Warren and I both realized that with the change in family finances, neither of us could rely on family help to attend college. If either of us wanted to get there, it would have to be on our own. I passed entrance exams for Brooklyn Tech, a vocational-technical high school in north Brooklyn, and started in January 1944—thirty-minutes from home on the BMT line, and a ten-minute walk at the far end. At the time, enrollment was 6800, all boys. Students were accepted from anywhere in the city. Classrooms were always full—thirty-five to forty students in every room—but overcrowded classes never seemed to be a problem.

During lunch, I worked as a busboy in the teacher's cafeteria. After school I was a clerk for TAMBE hobby shop, The American Model Builders Exchange. I worked counter sales, and built display models. Distributors would force storeowners to take slow moving merchandise in order to get more popular products. If customers saw what the model looked like they were more apt to buy the kit. Model building experience was helpful in my later career.

The storeowner used to buy model glue and paints—then called dope—in five-gallon cans. Next, we'd pick up cases of two, four and eight ounce clear bottles. Two of us would spend a Saturday pouring the strong smelling chemicals into the smaller containers and restock the shelves. At the time I was paid five dollars plus lunch to do that chore, and the first thing done was to open all of the windows to clear as much of the odor as possible. I wonder how much money today's paint sniffers would have paid the storeowner for that job.

Located across the street from Fort Greene Park, Tech is housed in an eight-story building crowned by a four-story central tower for radio station WNYC, the voice of New York City. Mayor Fiorello La Guardia's speeches to audiences in the 3,000-seat auditorium were broadcast over that station. The tower is the second tallest structure in Brooklyn; only the forty-two stories Williamsburg Bank building, located three blocks away, is higher.

Swimming at Coney Island and Brighton Beach was intermittent during the war years. Oil slicks from the tankers sunk off Sandy Hook coated the sand, leaving dark blotches where they washed in with the tide. Sun bathing was possible, but it was better to swim at Riis Park, Rockaway. A special bus route carried us down the southern length of Flatbush Avenue, across the bridge, and ended at the beach. Surf was higher there, and the undertow more severe, but it was great fun.

When I became old enough to obtain working papers, I worked Saturdays and summer vacations as a grocery clerk for Roulston's on the corner of Flatbush Avenue and Farragut Road; it was one of a chain. They were small stores with a manager, one or two clerks, and a delivery boy. Cigarettes were ten cents a pack during WW-II, and though not rationed they were often difficult to obtain. Some enterprising storeowners left loose cigarettes in a bowl and sold singles at a penny apiece. Smoking was an 'in' thing to do. It was easy to start, and took me twenty-five years to lose the habit.

Our cash registers at the store were mechanical, not the automated ones of today that tell you how much things cost and how much change is due. We shared one hand-cranked adding machine. If that was in use, the prices were written on the manila shopping bag and summed manually. In a largely immigrant neighborhood many of the ladies didn't speak English, but that didn't mean they didn't understand money. Woe to any clerk if they made as little as a penny error on the bill. What a tongue-lashing, "Yu tief! Yu crooooook! I teach yu to get rich on me!" And if they had anything in their hands, a purse or umbrella, the clerk could get hit on the head or wherever else they could reach, just to emphasize their words. It didn't come out in Kings English, but the meaning was clear. Arithmetic improved in a hurry.

In addition to money, we had to cope with ration coupons or red and blue plastic tokens known as points—they were required for meats, butter, sugar and other controlled items. I became skilled at estimating cuts from cheese wheels, or butter that came in barrels.

Whether a customer asked for a quarter pound or a pound, they didn't want it in three or four chunks.

By the time I was fifteen, I had decided that I'd teach Shop and Drafting at a vocational-technical school. With twelve courses to choose from at Brooklyn Tech, I selected the electrical program rather than college-preparatory. I felt that way I'd be prepared to work at a trade, attend college, and pay my own tuition.

Life wasn't all school and work for me; roller dance skating became my number one passion. Warren and I attended skating sessions at least twice each week at a roller-rink named Park Circle. It was located near the southwest entrance to Prospect Park in north Brooklyn, but we traveled to other skating rinks: Queens Rollerdrome, Mineola, Eastern Parkway and other places around New York City, Long Island, or New Jersey. Eventually I accumulated a hundred dollars, enough to have boots made by Stanzione and Sons; they worked from a loft building in Manhattan, creating custom-fitted boots for names including Sonja Henie.

Travel with my brother and his older circle of friends caused a small dilemma for mother and dad. They didn't want to throw a wet blanket on my social activities, but all of my female companions, including a skating partner, were older than I was. Four had finished school and were working. Dad sat me down for a bird, bees, and responsibility talk that made both his concern and position obvious. The gist of his message was, "Until you are much older, at least until you're finished high school, keep your hormones under control. Until you're ready to consider marriage, you will treat all girls as though they wore barbed wire. Look all you want, but don't touch." I was mature enough to know that I'd better take his words as serious, not as a joke.

In November of 1944 the family was informed that my cousin was dead. Lieutenant John Bishop had been killed in a plane crash at Guadalcanal, but it was years before all the details were known. His death put a crimp in my brother's educational plans. Warren understood the college benefits that he could obtain from the GI bill working its way through Congress, but enlistees under eighteen needed written parental consent. He had been trying to convince mom and dad to let him drop his senior year of high school and join the Navy. John's death prevented them from giving approval for months, but Warren pressured them even harder. Finally, in mid-1945, with the war ended in the European Theater, they gave reluctant consent. After basic training he volunteered for submarine duty. Stationed at Portsmouth, New Hampshire, he was assigned to the U.S.S. Clamagore.

I asked him once, "With your Brooklyn accent in the confines of a sub, did all the crew end up talking with your dialect?"

He laughed, and said, "No. One guy was from Alabama. Everyone wound up with a drawl."

Warren L.T. Bishop, USN
1945–1947

In February of 1946, I took part in a charity fundraiser at Madison Square Garden. It was a roller-skating review put on by city rinks in behalf of the Infantile Paralysis Foundation. After the demise of Polio, that group renamed themselves The March of Dimes, to become champions against birth defects. Park Circle rink had a theme built around Victor Herbert's music of Romany, or Gypsy, life. The costumes were colorful, and to perform in "The Garden" delivered an incredible feeling of accomplishment and satisfaction.

During the summer of 1946 I worked at Ferns Restaurant in Old Forge, New York, the heart of the Adirondack Mountains. First lake, in the Fulton chain of eight lakes, was only minutes away. With Mondays off and three hours between afternoon and evening shifts, there was ample time for swim, sun or fun. My employment required that I learn to drive an old pickup truck around the grounds, to do cleanup chores. What a summer, and what a difference to live in that mountainous, wooded place with three thousand people versus three million in Brooklyn. It was easy to become spoiled. When I returned home it took awhile before I settled back into the role of a Brooklynite; I found myself supercritical of dirt on subway station walls and streets.

Electrical classes at Brooklyn Tech taught me the properties of alternating current and direct current, or AC and DC. They covered power generation and proper handling of electrical hardware. In late 19th century there had been considerable debate as to the best power form to use. Nicolai Tesla tried to interest Thomas Edison in the use of AC. Edison formed a group that built the world's first power station, at Pearl Street, near the financial district in New York City. But it supplied DC, and power loss became too great after one-half mile. Tesla sold his patents to George Westinghouse, who built the first AC power station at Niagara Falls. Electricity was supplied to manufacturing plants in Buffalo, seventeen miles away. It was one of the rare times Edison chose the wrong side.

The electrical program that I entered at Brooklyn Tech was geared toward power distribution networks; electronics wasn't in our everyday vocabulary. In my last semester, the electrical class was called Radio Theory; my textbook was an RCA tube handbook. Television was an infant, picture tubes were miniscule, and transistors were unknown.

Computers such as Eniac were spoken of in reverence and awe. The machine occupied eighteen hundred square feet of space, weighed thirty tons, contained 18,000 vacuum tubes, and its prime function was to develop ballistic charts. To launch a projectile for a given range and target, it's necessary to know the exact elevation and azimuth of the gun barrel, cross wind direction and force, plus other possible conditions.

Students were told of Eniac's ability to work previously unsolvable problems, or tedious equations. It took hours to set up a problem and, despite tons of air conditioning, could blow a tube in minutes. However, it was state-of-the-art. Reverence had its limit for radio comics though; one joked that the lights of Philadelphia dimmed when the computer was in use. I've seen lights in a computer room go dim, but never thought to look outside. It was decades before solid-state transistors came into usage.

In my high school math classrooms, ten-foot long demonstrator slide rules hung above the blackboard, and I learned to be fluent with a ten-inch "slip-stick." Mine was a log-log-duplex-desitrig unit, with a magnifying lens over the cross hair. I received a discount on that model from my afternoon employer, and he spread repayment over a month. In many ways a slide rule was superior to the expensive Friden and Monroe mechanical calculators of the day because square roots could be found without the use of a special math table, as the machines required. One friend of mine loved to feed the calculator problems whose answer would be infinity, or a repeating decimal such as 1.3333. The machine would bounce around on the desk, and produce loud, grinding noises. Today any cheap, hand-held, battery powered, electronic calculator outperforms them all—without the vibrations and sound effects.

A catch-22 dilemma developed about my working in the electrical trade. In New York, no one works without a union card, and no one gets a card unless a blood relative is already in the union. I met twins who were visiting from the Panama Canal Zone; large contractors in the Zone trained their own workers. If I were became an electrician working for one of those companies the union would have to recognize my card when I returned. But I wasn't sure that moving to the Zone was best for me. I held off doing more than writing one company for their hiring requirements and details of the training program.

Ted Lyons, an electrical instructor at Brooklyn Tech had completed an apprenticeship with General Electric at Schenectady, New York, before he decided to teach. In my senior year, he brought a speaker to New York, to tell students about job opportunities at GE. Schenectady had filled their apprentice quota, but Massachusetts had changed their two-semester school year to a single year with only September entry. Transformer Division, in Pittsfield, had no January graduates to hire.

Carl Beers, a manager in the Pittsfield personnel department, presented slides of the plant and community. Short, portly, and dark-haired, he had an outgoing personality plus a booming voice. He was jovial, and his eyes twinkled like Santa Claus. But when he looked into my eyes, his stare was so intense I felt that he could see inside.

Pittsfield is the county seat for western Massachusetts. Located in the heart of the Berkshire Hills it is one hundred and fifty miles north of New York City, and fifty-eight miles east of Albany. Carl spoke to the class about post-war opportunities in the stable and growing industry of power generation and distribution. Low cost-of-living in that area made it attractive. Openings were available for apprentices to become a draftsman, toolmaker, electrician, or machine operator. College-level evening classes would be attended.

New York University offered an Industrial Arts program. I'd announced my decision to the family that I wanted to live at home and work jobs that didn't interfere with class. Mr. Lyons encouraged his students to take the GE entrance exams, if only for the experience. Ten of us did so. Months passed, and graduation was imminent. That brief contact with General Electric slipped from my memory.

6

"When handed lemons, make lemonade."

—Author unknown

Volunteering for the Army or Navy Reserve Officer Training Corps, or ROTC, was one idea of how to seek help toward college. They paid tuition, and a monthly stipend. I looked into it, but their requirements called for 20/40 vision uncorrected by glasses. Mine was 20/50. National Guard also had a vision requirement. Enough volunteering," I mused, "if they ever want me, they can draft me. I'll work full time until fall, and then at any job which doesn't interfere with classes at NYU."

Seymour Cohen, an earlier graduate of Brooklyn Tech, returned at intervals to recruit for Kessell Silver Plating Company. It was a family business that he operated for his father-in-law. Their shop was on the eighth floor of an old, loft style, factory building, mid-way between the Brooklyn and Manhattan Bridges. Kessell manufactured four lines of silver plate: trays, coffee and teapots, sugar and creamer sets, soup tureens, and the like. Hired at seventy-five cents per hour, my job was as a handyman: cast lead handles or spouts, shear trays to size, remove sharp burrs or edges, and run errands.

Brooklyn Tech had trained me to work on old machinery as well as new, but Kessell owned some of the most antiquated equipment I'd ever seen. If there had been any safety guards installed, they had been removed and lost. It was a time before OSHA, the Occupational Safety and Health Agency. Those machines were legal at the time. Protest, and there would be no work. The factory was dark and dismal in appearance. Glass in the windows had been painted over, and years of inattention or dirt didn't improve the light transmission. Numerous light fixtures over the aisles had no bulbs, though each machine did have a gooseneck lamp that could be flexed to aim at the work area.

I enjoyed any opportunity to go outside on errands, just to see the sunshine again. For a half-hour before lunch, I would take sandwich and drink orders for those who hadn't brought a brown bag; then I'd head out to the Italian deli on the corner. A hero sandwich and any bottled soft drink cost thirty-five cents, if you returned yesterday's bottle for the deposit.

At least once a week I went to a tiny jewelry store on Canal Street in lower Manhattan. On those trips, I'd pick up a moderately heavy package; it was always wrapped in a brown paper bag tied with string. I'd made four trips before learning that I was carrying two silver anodes for the plating bath. Mr. Kessell didn't spend money until it became necessary. When anodes eroded in the plating bath the plating action slowed or ceased. That was as soon as he'd buy the next pair. The man couldn't be described as frugal—he was a long way beyond frugal.

The only company phone was in the office, and one day I received a call from mother. She felt I'd want to know about a job offer I'd received from General Electric. When I took the call, Kessell glared at me. I cut the conversation short. He reached the door before I did, waggled a finger in my face, and said, "You don't get no more calls at voik. Not ven I'm paying you time."

All I responded to his verbal tirade was, "Yes, sir."

That afternoon, one of the machines acted up; it was an ancient type of shear. If the treadle were depressed on the left side, the shear cut blade cut toward the left stop, and vice-versa for the right side. The blade descended wrong, snagged the copper tray, and yanked it out of my bare hands. Those edges were as sharp as a knife blade. I opened my fingers, let go, and stepped back. The tray whirled past my stomach and sliced through my denim shop apron. As the tray jammed, the shear continued its descent. Two halves of the tray hit the floor and rang like a bell. I looked at the gash in my apron, and checked that it hadn't cut deeper. My hands trembled.

Kessell spied on his employees. He came out of the shadows, shook his fist at me and bellowed, "Dot tray cost me t'ree dollars. I'll tek it out of your pay."

So far, I'd bitten back my responses to him. Knowing I had another job, even if it was out of town, I shouted back. "I won't lose a hand for you. Don't expect me to give you blood for what you pay." With adrenaline flowing I felt self-righteous, but the need to explain to my family—the feeling that somehow I had let them down—that bothered me all the way home.

When I told the family, they laughed at my description of the encounter. Dad promptly lifted my self-imposed guilt. He shrugged, and mimicked Popeye's line in the funny papers, "Ya can stans so much, and ya can't stans no more."

My father saw life with remarkable simplicity. His philosophy was that a man could accomplish anything if he wanted it badly enough. We discussed the GE offer, and the more we talked the better it sounded. Starting pay of ninety-five cents an hour was more than Kessell or others in the city paid unskilled workers to start, and the offer included a raise every six months.

Finally he suggested, "You enjoy the mountains. Why not take their job offer for now. You'll have six months to make your next decision. Either come home before NYU classes start, or stay to complete the apprenticeship and save money awhile longer."

I agreed it sounded a reasonable thing to do, and would buy me time to make a final choice. A phone call to Carl Beers confirmed my acceptance. Two days later I rode the subway to 42nd Street in Manhattan, and changed for the cross-town shuttle to Grand Central station. I didn't realize it at the time, but except for a few weekend visits I was leaving home forever.

My afternoon was spent finding bus routes in Pittsfield, the personnel office where I was to report, and an inexpensive restaurant. Bridge Diner was a typical greasy spoon, built over the New York Central train tracks, but only one block from the YMCA where I'd reserved a room. On Leap Year Day, February 29, 1948, I reported to the Personnel office where they set me to filling out forms.

Carl Beers came over to greet me. He said, "I guess wide-open spaces and fresh air don't appeal to everyone. We made six offers to students in New York City. You're the only one that has accepted—at least so far." Then he added, "You have one more form to sign. Since you aren't eighteen, we'll have to send this home for your parents approval."

It was written in Old English script and looked archaic. The form told what GE would do for me, what I was expected to do for them, length of the indenture period, pay schedule, time off to go to church…huh? But the line that got my rapt attention said, "Therefore, Eric J. Bishop is hereby indentured, body and soul, to the General Electric Company for a period of—" I looked up. "What is this stuff? Time off for church…body and soul? Is this a joke, or what?"

His face split into a big grin. "No joke, but I love to see the expression on a person's face when they see that. Massachusetts is a Commonwealth, and indenture laws go back hundreds of years. That's the standard form for Indentured Servants. I'm not sure how it would hold up in court today, but without signing it the company can't take you on as an apprentice. Not in this state."

Needless to say it wasn't a stumbling block for me. Sixty-four of us started the program that year. For a minimum of six months, apprentices work in the machine shop; later assignments are made for a specific career choice. Evening classes in technical subjects are the same for each student, and were held three nights a week. I found myself surrounded by returned veterans from WW-II. Two had been caught in the Battle of the Bulge during the winter of 1944; another, a British chap, had flown a Spitfire fighter. For years to come I was always 'the kid' to them.

GE is a master at fine-tuning the education of their employees. They conduct special programs for apprentices, managers, new engineers, and business trainees. At intervals they have members of those groups meet with upper level management to discuss training progress, goals, or current issues. Individuals are selected for extra assignments or committees. We'd discuss our training, and recommend solutions to specific problems.

Apprentices sat in on the General Manager's staff meetings, to hear management's side of production or labor problems. That was how I met Robert Paxton, who became the President and eventually CEO of General Electric Company. We met monthly with individuals like Louis Berger, Ray Smith, Bill Ginn and others who became Division General Managers.

Pittsfield was a unionized plant, with labor contract provisions for such things as layoffs. That affected us in that apprentices would be laid off in the same proportion as other workers. It didn't seem fair to me that an apprentice, someone committed to a long-term program, could be laid off and never picked up again when the other workers might be back in a month. During one downturn, our supervisors did have to pick several apprentices for termination. One chap, Potter, was a slow learner on the powered machines, but did good work on forming brackets and simple tasks. Our floor supervisor came out of the office just as Potter pulled down on the three-foot handle of an arbor press. It knocked the man out cold. When he saw that the supervisor was rousing, he said to no one in particular, "Tell him that this week it's me!" Stopping momentarily at the coat rack, Potter left.

When it came time to move out of the training room, I thought about which program to enter. Brooklyn Tech had been directing me toward becoming an electrician, but the problem of not having a blood relative in the union could still crop up as a stumbling block. Tool-making was my next thought, but before I started there another option opened.

Ralph Munn, the Manager of Apprentices, asked whether I'd be interested in switching to their Time Standards and Methods program. Today it would be called Industrial Engineering: factory layout and relocation task analysis, tool or fixture design and manufacture, personnel training, and cost estimating. After having been in Kessell's shop, and the apprentice training room, work in an office sounded great. I joined the group and performed the job for five and a half months, until a management decision terminated the program.

Two of us who had less than six months on the Methods program went back to our original assignments. I worked in the toolroom for six months, but it wasn't what I wanted now. My heart wasn't in it. I'd had the taste of a white-collar position, and knew there was another way to make a living—no need to do heavy lifting, the greasy jobs, or to go home with cut and sore fingers.

Fortunately, Ralph was easy to talk with for career advice. Near retirement age, he had a lifetime of experience to draw upon and was a great source of information. It had been his intervention that caused me to detour into the Methods program. I asked him whether a transfer to Drafting was possible, even though months of my apprenticeship had gone by. He reviewed my evening class grades and work performance ratings. They were high, and he felt that the other assignment still constituted good work experience for me. Also, I'd had three years of drafting at Brooklyn Tech; I wasn't a beginner at detailing piece parts.

Ralph laughed, and told me, "You're the only one in the program to ever start on this many career paths." But he was agreeable to my request.

It was another unexpected twist that—although GE was the largest manufacturer of electrical equipment in the world—to achieve highest rank as a detailer, designer, or engineer you had to be mechanical. Electrical details such as cable size, bushing size and separation distance, core and winding size, insulation shape and thickness were outlined in the Standard Practices manual. Charts provided details for the parts manufacture, dependent upon the voltage and current required.

Distribution Transformers are the small ones seen on telephone poles or half buried in the ground for towns where no above ground wiring is permitted. The Power Transformers that I worked on were giants—the size of equipment for Hoover Dam. They could weigh up to two hundred and fifty tons before oil coolant was added to the tank. Bridges and train tunnels limit the height and width of the transformers shipped. It wasn't uncommon for a unit to be shipped on twelve to fourteen railroad cars. Final installation for units that size was always done in the field.

The exteriors of Power Transformers were customized to have the minimum container volume, which lowered the need for more oil coolant. Thermometers, heat radiators, or cooling fans had to be mounted. Input voltage for such giants was as high as 280,000 volts and needed petticoat bushings up to twenty-five feet in length; the output at neighborhood substations was 11,000 volts or higher. Mechanical work was custom work; it paid better, and became another of my accepted career changes.

One night I stopped at the Bridge Diner for a late dinner after class, and picked up a copy of the Berkshire Evening Eagle. A familiar waiter behind the counter nodded to me, poured a cup of coffee, and slid it down the countertop in my direction. It stopped within inches of my hand, without spilling a drop.

"Good eye, Jerry." Knowing the daily specials by heart, I added, "Make it meatloaf tonight."

He nodded to let me know that he'd heard. My eye scanned the newspaper until a column headline caught my attention. It read, "June Graduation Scheduled for Pittsfield High School." Dinner arrived, and I put the paper away thinking, "I'll read that at home." While eating, my thoughts turned to the word—home. I'd moved from the "Y" to a furnished room, but it was the first time I'd thought of a furnished room as home. I was living with 56,000 people instead of three million, machinery was newer and safer, the pay was better, and I'd found it easy to adjust to small town living.

"Home," my thoughts ran. "Guess I've found one."

Ralph Munn had suggested that I might think about completing the apprentice course before going on to college. He said, "It will always count as job experience when you apply for a teaching position. And, if you have to work before a teaching assignment opens, you'll be prepared to wait. You'll always be able to work."

Twice a week, I played softball at the YMCA camp at Pontoosuc Lake. That was where I met my wife. She had come out to watch a boyfriend pitch, but by the end of summer we were dating—the first girl I'd ever asked out who was younger than I was. Movies or an occasional dance were all we could afford, and on most dates we'd just walk and talk. When the weather got cold, we retreated to Benny's Diner on South Street, for a bottomless cup of coffee. The diner was only a block from her home on Church Street.

We intended to wait until I'd completed the apprentice program, but loneliness, general circumstances, and two cold winters expedited plans for us. In February 1950 we were married. She was nineteen; I was eight months older. Mature beyond our ages, but those times were hard. We did start marriage with furnishings that I refer to as, "Early matrimony—some of his folks furniture and some of hers." My experience working with dad made me a dedicated do-it-yourselfer. We were given an old metal bed, and I modernized it by adding imitation leather Masonite panels to the head and footboards. With wood from orange crates and remnants of the paneling, I built two nightstands and a vanity. Those things lasted until we could afford better. Housing near the plant was reasonable, and living close saved bus fare. From a few office locations in that huge plant, I was able to walk home for lunch.

Nine classmates were married; five were veterans restarting interrupted lives; two had married high school sweethearts. Married or single, our class tended to commiserate, socialize, and cling to each other for moral and physical support. Six had children, three had cars, and none had spare cash. One unmarried chap, Richard Pell, lived on a farm. He brought bushel baskets of fruit or produce to share. Living through a winter by eating mostly turnips and potatoes donated to the common cause, it's easy to know what friendship means.

When the Korean War broke out, three of my apprentice friends were recalled by the Air Force. With the chance increasing that I could be drafted, my spouse and I decided to wait before having children. It was too late.

7

"When you reach for the stars you may not quite get one, but you won't come up with a handful of mud either."

—Leo Burnett

When my daughter, Sheryl, got too heavy to carry on a bus I bought my first car. It was six years old, a 1946 Chevrolet Club Coupe that cost six hundred dollars. A businessman's car, there was a long sloping trunk that made the back seat short-legged for an adult, but it fit a crib mattress perfectly. We didn't use baby sitters often; that car allowed us to attend the one-dollar-a-car nights at the local drive-in theater. Sherrie would play, then watch the show for a while, and go to sleep. The only time we had a problem was at *The Wizard of Oz*. When the wicked witch's guards threw their spear-like halberds at Toto, Sherrie let out a blood-curling scream. We had to leave.

Of the sixty-four apprentices who started together, twenty-three graduated as the class of October 1953; I was proud to be one of them. Military recall, work skills, grades in class, or leaving the program—no matter the reason—less than forty percent of us completed it. My first assignment as a journeyman detailer was with the Machine Design group that designed and built specialized machinery for use in the factory. A 100-ton capacity spreader for an overhead crane, test equipment, hand tools, special brazing equipment. Whatever a production line needed, we produced in-house. Transformer Division was also equipping two new plants located in other cities, so the work was varied and plentiful. Overtime was common.

Once we had a car, it was possible to take scenic rides outside of the city. On one trip, we discovered the quiet little town of Lee, Massachusetts—twelve miles southeast—five thousand people, with land around the houses. Where we were then living near the GE plant had no fenced yard and the house was close to the street, Sheryl was limited to playing on the front porch. We returned to Lee several times, began to look at available houses, and moved to our first house on Halloween Day, 1953.

Six months later, the car began to have real problems. It needed to either be replaced or have the engine rebuilt. A new car was beyond budget, and I'd never worked on cars before. Joe Lahart, the owner of the service station I'd been using, offered me garage space so I could do as much work as possible for myself. He charged me for parts, or labor when his mechanic assisted me, and the entire job cost seventy-eight dollars.

Joe also ran a private club for wealthy people. It was a thirty-two-room mansion in Lenox, Massachusetts. Many members were summer visitors to the Berkshires. They could rent rooms or suites, and hold private parties, luncheons or dinners. Knowing that I could always use extra money, Joe asked me to work at the club whenever he needed staff for a special event. I became bartender, busboy, waiter or wine steward on occasion. Besides money, the workers could split up left-over food when the event was over. Roast beef, chops, rolls, cake—even the unfinished wine in the opened bottles—all went home. Sometimes the items taken home were worth more than the money I was paid. I've been fortunate, when in need, to encounter people like Joe. He extended a helping hand with no other thought than that I do the same for others.

The U.S. Justice Department branded the United Electrical Workers union a communist dominated group. National union leaders swiftly organized the International Union of Electrical Workers to prevent losing representation in plants where U.E. held contracts. In Pittsfield, there were two locals: one for factory workers, another for those in offices.

Voices of the people in drafting had always been a minority among the 2,000 members of the office workers local; they seized the opportunity to organize a new local. A petition was circulated through the drafting rooms to have the U.S. Labor Relations Board schedule a vote to allow a choice for the detailers and designers to choose between IUE and a newcomer, American Federation of Technical Engineers. AFTE won. Pittsfield drafting became a separate employee group of 550 eligible members, and 500 willingly joined the new local. It became a strong, new voice for several years. Then people began to fall back into the habit of avoiding meetings, or not being available to hold an office. When I tried to get information from my area steward, he said, "I only attend when they're working on the contract. There isn't much going on right now." He couldn't answer the simplest of questions for me. At the next election, I ran against him and won.

After two years as Steward, I was elected Recording Secretary—because I also attended the monthly general-business meetings. There was a bare quorum present on the night when a new slate of officers had to be nominated. There weren't enough people available to fill all of the offices unless I would accept one. I did; belonging to a union isn't my first choice, but when one exists, and determines your earning potential—it pays to be involved.

The Cold War continued, and all branches of the military continued weapon development programs. In June 1955, GE Naval Ordnance Department sent out a request for personnel. NOD was located three miles east of Transformer Division. They had jobs that needed prompt filling, and offered trial periods for advancement. I volunteered, and began a six-month trial period for promotion from Detail-1 to Design-3.

In my entire working career, it was the only office where I had a window to the outside world—and what a window. On three sides of the room, sixteen feet of one and one-quarter inch thick clear plastic sheets stand between the vertical steel I-beams that support the flat roof. Those window-walls offered a panorama view of lush New England foliage. Perched on top of the radar antenna assembly building, the room had originally been used as a test-berth until the equipment became too massive to fit the elevator. Appropriately nick-named the penthouse, the room's natural lighting made it an unquestioned choice for conversion to a drafting room.

Detail drafters worked in the interior, at four-foot long drawing boards. Rank does have its privilege, and designers were seated at eight-foot long drawing boards around the perimeter. I remember the thrill that coursed through me when, upon promotion to Design-3, it was my turn to occupy one of the envied window positions.

I'd also moved to another position with our AFTE local, that of Corresponding Secretary. As such, I was on the four-man Negotiating and Grievance Committee, which met with local management when there were complaints by any drafting employee. The position carried super-seniority, which meant that I'd be the third from the last in drafting to ever be laid off.

On a few occasions I alternated with my manager, Phil Weissbrod, to drive to meetings at the main office. But GE followed the axiom, "Caesar's wife must be beyond reproach." Phil was ordered to stop riding into the plant with a union officer; it was bad for appearances, no matter how innocent. That was an interesting era to be involved with labor relations. The man in charge of all labor contracts for General Electric was Lemuel Boulware, a hard-nosed individual. Once he made an offer, he never budged. The press began to refer to any non-negotiable situation as Boulwarism. His position was later condemned by the U.S. Labor Department as "failure to negotiate in good faith."

Officers of our local gave input to the national AFTE contract team; periodically we met with officers from other locals in the GE chain. We alternated meeting locations for everyone's convenience, and discussed what we would like to see in the next contract, or compared what was different about local supplemental contracts; i.e., what did one group have in their contract that others didn't have in theirs. On one occasion it was noticed that we were the only local that still had the right to strike over unresolved grievances, and that clause was good for three years. Asked how we ever got GE to agree to that, our local President said, "We haven't abused the privilege by calling the wildcat walkouts that are so popular in the other plants." Two years later he moved to Washington, to work in the National AFTE office. His farewell dinner saw more than a few managers attend, and attest to their regret in losing his services.

It's a short distance from Russia to the United States across the top of the world. Radar was still a valuable tool in providing warnings of activity that could threaten national security. A string of large radar antennas known as the DEW system, or Distant Early Warning, was to be installed across Alaska and Canada. They watch the icy wastes at the top of North America, to warn against missile or bomber threat. The first antenna that I worked on, the FPS-7, had a 25 by 40-foot parabolic shaped bowl. It was designed in lightweight sections, to allow field assembly in the Arctic. That was followed by design of the drive base, or gearbox, which rotates and tilts the antenna. On one occasion, Navy officials wanted to try using a steel framework for a ship-mounted antenna, I think it was numbered SPS-8. It turned into a fiasco and the steel idea was quashed.

Mark 44 torpedo was next. It was the first acoustic or sound-seeker model. I designed an exercise section that programmed the unit's activity during a test; after its practice run, weight was jettisoned to make the torpedo buoyant for recovery. Its steering and propulsion design was the first to implement counter-rotating shafts to drive the two propellers. For testing, our units were shipped to Key West, Florida, where the torpedo was rolled into the ocean from the deck of an old fishing trawler. Seawater activated the batteries, and a timer gave the ship time to get away—except once. The timer failed, the battery came to life within seconds, and our exercise unit punched a hole through the side of the wooden trawler. It sank. GE is a self-insurer; they replaced the boat.

The Mark 56 Gun Director had been the backbone of the fleet for anti-aircraft defense during World War II and Korea, but it wasn't capable of tracking fighters at supersonic speed. Our newer design, the Mark 73, moved the gunner's position off the rotation mechanism; it turned so fast that any operator would be thrown out of their seat.

My first opportunity to work with an expatriate German engineer was in 1956. Helmut and I were members of a cost reduction team for MPQ-4, an Army field antenna system mounted on a truck chassis. The prototype for the leveling legs was cobbled together from welded aluminum plates; for the production run we had to decide upon an economical way to build them. Helmut made a suggestion. He said, "Form mirror-image right and left hand parts from sheet metal. Weld them together with one seam down the middle."

Detractors said, "A sheet metal leg will bow under the weight of the truck."

His defense was, "So what. The initial load bends it, but wind and other loads are so light that it won't deflect any further."

He won that debate. The part was produced at great savings, and I was impressed by the practical nature of the man. Later he mentioned,"V-2 components needed that kind of approach. There was a need for a pump that had to sit for an unknown time, but instantly function after years of waiting. We used an off-the-shelf fire pump. It met all of the requirements. The best solution is often a simple one."

Our cost reduction committee was the first in GE history to produce savings in excess of one million dollars on a production run of only one hundred trucks—saving ten thousand dollars per vehicle. Helmut told me later, "We could have done better with the drive mechanism if we had time. Because of the water immersion requirement it was made of stainless steel, but it would delay production to redesign it with hardened steel parts." Perhaps to see if I'd rise to his jest, he added, "Sometimes I wonder how you guys ever won the war."

Fall is the colorful season to be anywhere in New England. Color explodes onto the scene at first frost. That can occur any time after Labor Day. Crimson, rust, and gold break out like measles among the stately elms, maples, oak, hickory, or white birch—not to mention shrubbery. Evergreens and slash pines add brilliance to their needles. There isn't a bad day to see foliage; it's either good or outstanding. But despite living in that scenic a location, I was hungry for something that seemed impossible for me to achieve in Pittsfield—a college degree. Two of my apprentice classmates had left to attend college full-time. Married, with a family to provide for, I needed another solution. Allowing credit for my apprentice classes I'd be half way to my goal. Nonetheless, two years of day school was still the equivalent of four years at night.

I researched nearby colleges. Rensaleer Polytechnic Institute was fifty-six miles away, in New York State; they offered evening classes. Amherst was further away, and offered only a few classes in the evening. A coworker gave me the name of someone they knew who drove to Rensaleer. I called him for information on classes, and their carpool.

"There are two others driving with me," he said. "Let's all talk. We'll meet you at East Plant-3, at 5:15 p.m." There are only two buildings in the Ordnance or East Plant. It was the pub at the next crossroad, Timmy Shea's place, which was dubbed EP-3.

When they arrived, introductions were made all around. Then I asked, "Where do you start from and when?"

"Well—we don't work close enough to each other to leave from the plant, and not close enough to home to go there first. It's easier to meet at Park Square, and drop the spare cars there. GE permits us to leave at 3:30 p.m. on school nights. We get to the Park by 4:00, and leave by 4:15. If you miss that time, you drive solo."

"How late do you get back here?"

"About eleven p.m." He hesitated, and then added, "Unless the weather gets bad."

"What time then?"

The three of them looked at each other, deciding who would answer. Finally one of them got brave. "Oh—we put on chains. We're usually at Park Square by midnight."

It was like pulling teeth; I still didn't have a complete answer. "What about the exceptions that get you back past midnight?"

"We wait at the foot of the mountain for a snow plow to arrive, and follow it back."

Their solution needed more thinking before I'd dedicate four years of evenings to it. No matter what time the group arrived in Pittsfield, there would be another half-hour to reach South Lee where my family was now living. Eager as I was for a degree, that kind of travel or time didn't have me jumping with joy. There had to be another answer. If possible I wanted to find it before next fall.

Phil Weissbrod called me into his office. It was March 1957, and he wanted to discuss my latest performance appraisal. It was an excellent review from my perspective. Then he came to the last item, "What can the company do to further your career?"

We had talked before. Phil knew my long-term aspirations, and I'd talked repeatedly with my spouse about a degree. Any eventual move would affect her directly, and she had lived in or near Pittsfield all her life. It was becoming obvious to me that I'd have to relocate to attend school; I wanted her to understand that too. Without a degree I'd run into the money ceiling.

Relocation was a difficult choice for the entire family. South Lee was a community of 102 homes in a beautiful setting. On October Mountain, which faced our home, there was a forest site from the Revolutionary War. Gentleman Johnny Burgoyne, the British commander, moved his cannon along the edge of the mountain rather than the valley floor. Lining the path with rocks and tree trunks created a corduroy road, which can still be seen as a slash against the hillside greenery. Our farm had a dozen fruit trees, stone fences, a stocked brook, and spring water that the previous owner had thoughtfully piped to the kitchen sink.

It was situated one mile from Stockbridge Playhouse a summer home to famous actors doing stock theater, only two miles from Norman Rockwell's studio in Stockbridge, five miles from the summer home of the Boston Symphony Orchestra at Tanglewood. Sherrie attended school in a three-room schoolhouse, but there were so few children that each received very personalized attention. She couldn't wait to go each day. I'd had to convince all of us that a move was in our best interest. How could a move anywhere be an improvement on what we had?

South Lee, Massachusetts

But I sighed, and asked Phil, "Look into a transfer site for me. Where I can stay with the company, but close enough to an engineering school that I can attend at night."

Phil studied me for a minute, to see whether I was serious. He must have believed it. "I'll hate to lose you, but I understand your choices. Let's see what there is right now." His fingers rippled through a folder. He stopped, and looked up. "Interested in San Jose, California?" he asked. "Atomic Reactors Division wants mechanical designers."

"California would be a hard sell at home, Phil. Three thousand miles is a long way from both families. If we fly back and forth to visit family, that could get expensive. Does it list places along the East Coast?"

His fingers rippled through a few pages. "How about Philadelphia," he ventured, "that's three hundred miles from here. Missile and Space Department is only months old; they need all kinds of people." He pulled out another sheet and read it. Here's a group that needs structural, packaging, and machine design." Scanning down the page he added, "Hell, I think they had you in mind when they wrote this list—must be ten jobs you qualify for."

"I meant to ask you earlier, Phil. Does my request to look elsewhere obligate me in any way."

"None at all."

"Can I borrow that Philadelphia list? Just overnight—I'll need to talk about this opportunity with my wife."

"That's understandable," he replied, and handed the papers to me. "Copy the M&SD pages and give me back the originals."

Later that evening, I read the general information about that location. Though presently located at 32nd and Chestnut Streets in Philadelphia, the division owned a hundred and ten acres in King of Prussia, twenty-six miles northwest. Ground wouldn't be broken for new buildings there until growth was justified. K of P is adjacent to Valley Forge National Park, and named for a Prussian King who toured the town and park area sometime after the Revolutionary War.

Phil also did research that night. Next morning, he told me that the GE building in Philly was directly across the street from Drexel Institute of Technology. During the day he telephoned the M&SD personnel office, and spoke with them. The next weekend I drove to Philadelphia for interviews. My family went with me, to look over the general area and housing. Four interviews had been set up before I arrived. After listening to the managers, I decided that Instrumentation group was the best choice for me. Phil Weissbrod asked that I remain three weeks in Pittsfield, to finish my present assignment.

I'm an avid reader, a sci-fi buff, and a member of "First Fandom." That's a group of people who began to read fiction in the early days of pulp magazines and paperbacks. The club logo calls us, "The dinosaurs of science fiction." To work on programs others might only read about added a thrill for me.

Our farm was put up for sale and I commuted six weeks until it sold. While I'd owned the place, I had torn down the barn and built a two-car garage with a long breezeway that connected it to the house. My property taxes were only one hundred and forty nine dollars a year for the house and seven acres. I hadn't wanted to upset that low tax by causing a reappraisal; before I started the reconstruction I'd reviewed building codes. The inside surface of the breezeway was covered with wainscoting salvaged from another house, but not along the top surface tying the garage to the main building. That section was left open until March 1957, when the county appraiser came by for a look.

To save him time I said, "Let me show you what I've done. The code says I can tear down and rebuild anything without a reappraisal, as long as by the first of the next year it's reconnected to the house by a single stick of lumber. There it is."

Barn Replacement, Breezeway and Garage

He looked up at the exposed beam, then at me, and burst out laughing. "A single stick of lumber. Huh! That thing is at least twenty feet long."

"Twenty four to be exact."

"Well, treat me to a cup of coffee" he said. "If you found that loophole there's no sense freezing my buns off out here while I look for something else."

Taxes stayed the same.

When the house did sell, my realtor called the existing mortgage holder to arrange new financing. That proved to be simple because Lee was such a small town. The realtor called Burt Nettleton, president of the bank that held my existing mortgage. The banker said, "Put your seller on the phone." When I took it, he asked, "What changes have you made?" I told him. When the realtor got back on the phone, the banker said, "It's okay, and you won't need an inspection. That house was my grandmother's. I know everything else about the place."

A former neighbor wrote to tell me a final story about that house. When I'd built the garage I'd been short of cash to buy a new garage door, so I used the old barn doors for the time being, but moved before I had a chance to replace them. The community firehouse had to be renovated, and they sought a temporary place to keep the pumper truck. Mine was the only garage near the center of town with a door tall enough to accommodate it.

There were others who found 1957 a year in which a move became necessary. "Da Bums," the Brooklyn Dodgers struck out in negotiations for a new stadium, or a satisfactory lease renewal. They moved to Los Angeles, California and ended decades of legend making at Ebbett's Field. The first years of the Dodger shift didn't go any better than mine. Disappointment number one for me was that someone else had filled the job in the Instrument group before I reported for work. Sent by the Personnel office to my second choice, Arming and Fusing Section, I found that the manager with whom I'd interviewed had already moved to another position.

Young, and naive, perhaps because of my desire to be where I could attend college, or simple trust in fairness, I overlooked bad signs. We rented a town house in Drexel Hill, west of the city, to wait and see where the best place to live would be once the plant moved. Cost-of-living was high in the Philadelphia region. My offer letter stated that, in order to compensate for the higher cost-of-living, I would be reviewed for salary action at six months and one year after my arrival. The job level was also scheduled for increase in six months to Design-2.

Army missile programs were all I'd ever worked on in that office, none of the space projects I'd envisioned. One day it dawned on me—like waking from a dream—everything I'd designed these last months was to insure that something would blow up. I'm not a pacifist, but the thought didn't set well with me. When I applied to Drexel, the other shoe dropped—they couldn't allow credit for the fourteen apprentice classes. I was floored at the thought that it wouldn't be four years of evening school, but eight.

Because of my time as a detailer and designer, Drexel allowed me to test out for three classes: Engineering Graphics I, II, and III. I attended the final exam for the highest-level course; passing that test meant they'd credit all three. The material covered was essentially Descriptive Geometry. Experience had me past that level of drawing. I finished in forty-five minutes, but the rule for final exams was that you had to be present for one-hour minimum. I sat and reviewed my work. The professor saw me turning the pages and motioned me forward.

"If you want to wait, I'll grade it now."

He took my papers and I returned to my desk. A few minutes later he looked up again and motioned. The grade was written in red ink—eighty-five. I flipped each page and scanned it to see which problems I'd made errors on.

He looked up again and asked, "Is something wrong?"

"My pride is wounded. I thought I'd done better. I'm looking for the mistakes."

"You didn't make any," he answered.

"Well I saw the eighty-five and—"

He was now speaking loud enough for the entire class to hear. "Don't knock that grade son. It's the highest anyone in this room will receive tonight."

You could read the faces of his students; they were in shock. So was I. The grade didn't strike me as fair; I wanted to achieve the best weighted-average possible. He was throwing his authority around, but there was no one to appeal to—this man was the department head. Wind was knocked out of my sails again; I'd passed, but I left in a somber mood.

Six months after I'd moved to Philadelphia, my appraisal brought a token raise but no level change. When I inquired, the manager informed me that, "I have two others that need their level increased to Design-2, and they were here before you. Personnel should never promise what can't be delivered."

"The letter was written by your predecessor, and sent as an honest offer; it went through Personnel for the record. Are you saying that you won't honor it?"

He answered, "Not right now. You aren't here long enough. There are others who need to have their levels brought up first." I knew who the others were. They played cards with this man every lunch hour.

Trying one more time I said, "It isn't a matter of time in this office, it's a matter of job experience."

His mouth curled into a sneer. He said, "Sue me," and walked away.

Nothing about the move had gone well. My wife had been ill for a month; our paycheck barely met the bills in this area, and now this. But it was not a good time for me to move again. I did interview with other local companies, but didn't jump at what was offered. Raytheon and RCA were hiring, but they brought in "new hires" at eighty percent of top pay so there would be room for salary improvement—a reasonable approach, but not to my advantage. And, I'd be back in a position of having to prove my skills all over again—back to square one.

A manager from Electric Boat Company in Groton, Connecticut, was doing interviews in Philadelphia; I talked with him. His offer was a lateral transfer—same salary, and they'd pay my relocation costs. That would at least take my family out of this high cost of living area. I told them I'd like to think it over for a day or two.

We decided not to jump too soon. I'd hang in with General Electric for now. I'm not sure what I expected to see happen, but something had to change. It sure did. I was rushed to the University of Pennsylvania Hospital for an emergency appendectomy; they burst while in I was in surgery. My doctor told me to stay home for two weeks, and GE put me on extended medical leave.

8

World War-II was twelve years past. Thoughts of Germans having been an enemy faded in the United States. More of the expatriates transferred to Huntsville Arsenal, in Alabama, and once again under Von Braun's leadership they built rockets. Now it was for the U.S. Army Workers involved with the defense industry jokingly referred to Huntsville as Hunsville, or Peenemuende South. At that time, the most powerful U.S. rocket was Jupiter with 70,000 pounds of thrust. Redstone was in development.

World scientists chose that time to appeal for their colleagues to make a united effort in learning more about the planet we live on, and the heavens that surround us. They established International Geophysical Year, an eighteen-month period that overlapped part of 1957 and 1958. It was dedicated to the coordination of specific research tasks.

In 1957, the Army proposed a satellite launch vehicle to a congressional committee: first stage Redstone, a rotating cluster of Sergeant rockets for the second stage, and a final Sergeant to inject the payload into orbit. Service rivalry caused the committee to shelve the Army proposal in favor of Navy's Vanguard. After the December 6 failure, when the Vanguard missile collapsed and exploded, Army renewed its proposal to Congress. On January 30, 1958 the United States orbited an Explorer-I satellite. Its payload of 18.13 pounds contained two radios, two antennas, batteries, and eleven pounds of scientific equipment. But Explorer-II failed to achieve orbit when the second stage sergeants failed to ignite.

Vanguard experienced another failure. On March 17 the U.S. achieved its next success with the Navy's third rocket. A small payload, six inches in diameter and a mere 3.25 pounds, the first Vanguard launched satellite confirmed an equatorial bulge caused by earth's rotation. For the first time, science measured the pear shape of earth. Explorer-III satellite followed on March 26. Despite the USSR launch of Sputnik on October 4, 1957, and Laika on November 3, discovery of the Van Allen radiation belt by Explorer-I, and mapping of the belt's shape from Explorer-III, was recognized as the greatest scientific accomplishments of the International Geophysical Year.

1958 proved to be the turning point for both the United States space program and my part in it. In the wake of those successful satellite launches, President Dwight Eisenhower announced that the charter for the National Committee for Aeronautics was redrawn to create the National Aeronautics and Space Administration. Eisenhower surprised many by insisting that a civilian agency coordinate space activities, but he was right. Elimination or curtailing of the paralyzing service rivalry was a major step to sending the U.S. into space.

As for me, I had an unexpected contact while recuperating from surgery, early in 1958. Someone in the Personnel office called me at home. At first my suspicious mind wondered if they were checking that I was really ill and homebound, but that wasn't the case. They inquired as to when I'd be available to interview with Howard Wittner. I'd never heard of him and asked for more information.

"Mr. Wittner is organizing a unit called Advanced Engineering. The department is being deluged with studies and proposals. It interferes with work routine in some groups, but the company doesn't want to decline any bids. He's asked us to locate five designers for him. Is there any way you could come in this week—or even tomorrow?"

It interested me that they were pressing me to come in soon. Working on studies and proposals sure sounded better than what I'd been doing in the Arming and Fusing group.

The clerk went on to tell me, "Mr. Wittner has obtained approval for a new position called Engineering Designer. It's for non-degreed persons who can do their own supporting calculations and work independently." There was already a position called Engineering Aide. I didn't know the difference, but thought it would be interesting to find out. Once they were assured that I wouldn't be in medical harm to come in for an hour, they made a ten o'clock appointment for the next morning.

Howard's secretary announced me. He rose and offered his hand over the table. "Sit down, please." For what seemed an eternity, he looked straight at me. His eyes carried intense feeling. Under that thoughtful scrutiny, I wondered whether it was meant to intimidate. It didn't. I had nothing to lose, and was prepared to relocate to another job or company if need be. That is, if I couldn't work something out here.

"I've been looking to hire five people," Howard said. "Personnel has a list of education and experience that I need. They are comparing that against the information in their Engineering Register. Bill Schoble's name came up and I hired him. We modified the parameters I'd asked for and found Ed Williams; he'll be on board next Monday. Changing those requirements again gave me your name—just three people I've found so far. Advanced Engineering needs people with your background. But first, may I ask a personal question?"

"Yes, sir. I'll answer if I can."

"Does somebody have something against you? Your salary and level are unusually low for the experience and education in your folder."

Now that was one for the book. Managers usually fence words about why they can't give money or another level. This man wanted to know why I didn't have both of those things. I thought, "He's either honest, or tricky."

"The manager whose offer I accepted last year wasn't here when I arrived," I told him. "His replacement hasn't honored the conditions that I'd agreed to. He's playing word games, twisting the terms in the offer and acceptance letters. Personnel can't, or won't do anything. I've interviewed elsewhere, and I'm one step away from leaving."

"That's straight enough, but you have ten years service with GE. It would be a shame to throw that seniority away. Let's talk, and see what can be done."

Howard inquired about my previous work, the apprentice classes, and asked for details of my last performance appraisal. He told me of the heavy workload that the A/E group already had on hand, and what appeared to be in store for the future. Finally he said, "Your performance review was less than six months ago—that's too recent for me to change anything today. My hands are tied."

"Well, if you can't help," I said, "it's time for me to decide whether I'll move on." I started to rise.

Howard held up his hand. "I said I can't do it today. I'm hurting to get people on board fast enough to meet the load. We'll be working twenty hours of overtime every week for months. That should relieve your immediate money problem. You believe in your ability. Show me what you can do. Give it six months and we'll both know what your services are worth. Then we can get your salary and level in shape. Will you trust that I mean what I say?"

His challenge sounded sincere. He offered his hand. Timing for a move was terrible, and college was closer here than it would be in Connecticut. "It's a deal," I replied, and we shook hands.

My surgeon let me return to work the next week, as long as I'd be on light duty. Howard cleared the way for me to leave my previous assignment immediately. Monday, I went in to my old office at 7:00 a.m., packed my equipment, and left before that manager arrived. No fourth or fifth designer ever came to work in A/E until years later, by which time the Missile and Space Department had grown to a Division. I'm not sure whether Howard quit looking, or Personnel never found others that survived his scrutiny.

All U.S. military branches continued to develop offensive and defensive weapons to counter the Soviet threat. The United States Air Force conducted high-altitude hydrogen bomb tests on July 31, and

August 10, from Johnston Island in the Pacific area. On August 27, 1958 a missile was successfully launched from a Polaris submarine.

Every branch of Advanced Engineering worked short-handed. Howard was correct when he predicted overtime; it was constant, and cured my immediate money crunch. Between work and school it was difficult to have time with the family. Someone once told me once, "The trouble with overtime is that when you have time you don't have the money, and when you have money you don't have the time." True—all too true.

I worked a two-week assignment near the shock-test laboratory, where scale models of the proposed reentry shapes were tested. That lab was located in an old taxi garage on D Street. A six-inch diameter tube was mounted horizontally as the gas cannon. Two-inch diameter shells or projectiles, of the shape being tested, were wrapped in a foam casing known by the French name of sabot. At one end of the tube was a pressure chamber; the other stopped in a concrete bunker, which was sandbagged for safety. Pressure built up until a diaphragm would rupture, causing the shell and sabot to launch. Partway down the tube, a stripper removed the sabot, and the shell flew past a series of cameras to record shock waves produced by the different shapes.

A Klaxon horn sounded one minute before every firing—to alert the workers. Everyone hurried to pull covers over their desks and drawing boards. Thump. The familiar sound indicated that the projectile was launched. An increasing whistle followed the progress of the shell in the tube, then a dull thud when the shell landed in the sandbag chamber. This was the moment we'd prepared for. We weren't in danger from the projectiles; warning was given because of the dust and dirt. The concussion released years of dirt particles from their slight hold on exposed metal rafters; a haze formed above us, and dust dribbled down like a mist. We averted our eyes. When the air cleared the covers could be removed again, it would be an hour or more before another shock wave test could be performed.

Wright Air Development Center, WADC, had been working on Project 7969 or manned space capsule for a year, and Air Force suggested a catch-up program called by the acronym MISS, for Man In Space Soonest, to launch an American as the first human being to enter space. They proposed that a modified X-series rocket plane be launched by Army's Jupiter rocket. NASA agreed to review the concept. X-15 was closely identified with the Air Force. WADC wanted data developed for an alternate vehicle, in case one was required, and wanted the data to be developed by a non-military source. GE Missile and Space Department received a contract to perform a ninety-day study of the capsule that WADC considered feasible for manned space flight.

I'd worked small individual projects after I joined Advanced Engineering, but when the MISS study was assigned to us it required half of our staff. For the vehicle, WADC defined a capsule shape with a slow reentry velocity and maximum volume for an occupant. The front to rear cross-section was a spherical nose, a cylinder in the center, ended by another cone portion. It looked like a bullet with a flared skirt at the rear—a stabilized cannon ball. In his book, Jules Verne didn't give enough thought to an air supply so his passengers could breath, nor was he correct in assuming that a porthole could be opened to the vacuum of space. But the longer I looked at the proposed shape, the more I felt that he'd hit it pretty close when he wrote about use of an artillery shell.

An occupant faced forward at launch, and on their back at reentry; the seat needed to swivel. GE was to provide a full-size mockup, and estimate cost. We could have met the basic requirements to study only the proposed shape, but our team of consultants felt other capsules should be proposed for the mission—either for improved accommodation of the occupant, or cost. Howard convinced GE management that this was an opportunity to become a prime contractor on what could develop into a major program. He requested and received permission to approve up to forty hours of overtime per week for key people.

Technicians were scheduled on two twelve-hour shifts, to allow work to be continuous, and without being in each other's way. The rest of us worked long hours at the beginning or end of the normal day, to always have someone available to provide direction.

In late September, Air Force staff arrived for an oral and visual program ending the study. Our night crew had barely finished last minute details, but there were three eight-foot diameter mockups for review. The requested shape had a functional chair; it rotated, and had other functioning hardware. Two stationary models showed other ideas that the GE team considered noteworthy. There was a formal presentation of several books outlining the capsule details, cost, et al; the bulk of the copies in that report had been shipped to WADC. The officer-in-charge praised the quality of our presentation, and they took all study data under advisement, but that was the last we heard from the Air Force on manned space capsules.

At the same time that we were performing the WADC study, NASA held their review of the MISS concept using a modified X-15. Jupiter with 70,000 pounds of thrust wasn't adequate to lift the vehicle's weight. A cluster of rockets was considered likely, and ideas ran to three newer, but untested rockets to lift an X-15B. NASA's mission was not to just do something, but to create and support long-range plans. Though suitable for a ballistic flight, the X-15 wasn't what would be needed for orbital flights. By the time the rocket changed to a Redstone, and a manned capsule substituted for the airplane cockpit, the entire system was different. Project Mercury—a single occupant capsule—was revealed to the world in the fall of 1958. Bids would be sought from contractors.

No pilot training or experience was called for on the first civil service job specification for a Mercury astronaut. President Eisenhower stepped in. He feared that loose requirements would allow millions of people to apply, paralyzing the effort. NASA was ordered to select from the ranks of military test pilots. That limited the list to five hundred and forty candidates. Physical and mental requirements were established. Unfortunately, the six-foot diameter of a Mercury capsule limited height and weight. That eliminated many pilots, including all of those who had flown the X-1 through X-15 rocket planes fifty miles high—closer to the fringes of space than anyone else.

The requirement imposed by the President, that applicants must have experience as test pilots, eliminated women. There were many that flew airplanes, but none were test pilots. A private medical study of numerous female pilots certified that a dozen ladies met all of the other requirements. Nonetheless, NASA didn't budge on their ruling.

"Spam in a can," was a derogatory term used by candidates, to express their feelings that the occupant would be a well-qualified pilot with no control over the vehicle in flight. When chimpanzees were proposed for several flights before the astronauts would be launched, the label became "Chimp mode."

General Electric is not an airframe manufacturer, and they entered the Project Mercury bidding as a partner to North American Aircraft. Several other companies also bid with one or more partners who could bring knowledge or experience in special areas. GE's role was to design life support, experiments, computerization, and heat shield. North American would design the capsule airframe, parachute system, and all other flight or recovery gear. The two design teams had to work in close proximity; it was simpler to send our team to work at NAA's headquarters in California. Twenty-four people from Advanced Engineering flew there, and put in twelve-hour workdays coordinating ideas with the NAA staff.

I'd worked in large buildings, but North American's use of an old hangar as office space was impressive. In this day of CAD, or Computer Assisted Design, few engineers work on a drawing board; many never even see one in an office today. But in 1958 all aircraft engineers and designers performed their work at one. When I looked in front of me, all I could see in that huge building was rows of boards with people hunched over them. Through the years, older draftsmen had told me told stories about having worked where "All you could see were ass-holes and elbows." This was a first for me. Now I'm a believer.

Four drawing boards were left out in the center of the room. A single desk and two file cabinets stood there. That space, I soon learned, was the domain of North American's Chief Engineer. Howard introduced his people. The Chief scanned the group. Then he turned back to Howard and asked, "Now that I've met your talking engineers, when do I meet your working engineers?" The question caught us by surprise. In looking around it was apparent that the NAA people were dressed in short sleeve shirts or had their sleeves rolled up—there wasn't a jacket on anyone in sight—except us. One by one ours came off. Red-faced, Howard also removed his tailored suit coat. The Chief laughed, and said, "Pick out boards for yourselves—any that don't already have a body already leaning on it."

As if the Chief wasn't enough of a character to cope with, NAA assigned a cost analyst to us—Stanley. His job was to review all dollar figures that GE's staff came up with before those numbers were accepted as final. He was quiet and reserved, but recognized the power of his position. No number got past Stanley without discussion, and it was inevitable that he'd adjust them. One evening near the end of the project, he built up a head of steam. About 8:00 p.m. we heard him grunt, get out of his chair, and begin to pace, Stanley began to mutter, we could only make out a word here or there. He glared at the papers in his left hand, then those in his right, and flung the offending papers in the air.

"Fifty thousand," he exclaimed. "A hundred thousand. How the f—am I supposed to know what it costs to flight condition a goddamn ape?"

There was dead silence for a few seconds. Then, someone applauded and yelled "Attaboy, Stanley." Another cried out, "You tell 'em, Stanley." The rest joined in—clapping and cheering. He sat down, embarrassed for his profane outburst in front of a dozen people.

The schedule pressure was getting to us all. We needed a break, at least for one night. It wasn't hard to talk everyone into going for a nightcap. At the club, he told us to call him Stan, and he was a great guy to party with, but the next day, <u>Stanley</u> was the personality that returned to work.

Thirty-five pilots were asked to participate in tests to find the final crews for Mercury. A committee of medical personnel made the final decisions. Early in 1959, T. Keith Glennon, NASA's first Executive Director, introduced seven Mercury astronauts to an anxious world. New training methods were developed for tasks the astronauts would perform under conditions that no humans had encountered before. The first of the training programs that needed development was to teach astronauts to function in weightlessness.

One way is to fly a geometric path known as a parabola; NASA calls it by the fancy name of Keplerian trajectory. When the specially equipped, padded airplane comes over the top of the arc and begins to descend, there is a period of zero gravity—one to two minutes. Airplanes are the first training sites where humans are introduced to space activities. The duration time is short, but they fly the trajectory over and over.

Another method is to ballast trainees to be neutrally buoyant in water, to neither rise nor fall. Astronaut movements are slow and deliberate. As long as the underwater trainee moves slowly, and the structures are modified to minimize or eliminate water drag, the training is realistic. Tasks can be performed in water for hours at a time. If the work proves difficult, or a tool inadequate, the task or tool is modified and testing is repeated.

It was also necessary to determine limits that are the reverse of weightlessness. To determine the mechanical stress limits with which to design rocket engines, fuel tanks, and structures that would carry humans into space, NASA had to determine what the human body could tolerate at launch and reentry. Each 'g' is equal to the force of gravity. Rocket sled tests were performed at White Sands, New Mexico, with animals and humans in a simulated cockpit. Other tests were at Navy's human centrifuge facility in Johnsville, Pennsylvania, where the gondola is mounted at the end of a fifty-foot arm. A pilot could be subject to any desired force up to forty g's.

The Air Force wasn't totally bypassed by NASA's takeover of space programs. They continued experiments with rocket propelled X-planes at Edwards Air Force base, began design of classified satellites for military communication or other purposes, and planned missions such as the Manned Orbiting Laboratory.

With Drexel Institute located across the street, it was possible to work until ten minutes before class and still arrive on time. Occasionally my work schedule was so tight that I returned to the office after class. The Advanced Engineering group expanded, and middle-level managers were delegated over the new units. After the Mercury bid I rarely saw Howard, and didn't take assignments directly from him now.

The six-months we had talked of came and went. It became seven, then eight and nine. He was traveling a lot, and our paths didn't cross. Overtime pay was filling the cash needs at home, but I began to wonder if I'd been lied to—again.

9

"I know God won't give me anything I can't handle. I just wish he didn't trust me so much."

—Mother Theresa

The last workday before Christmas was payday. Howard came over to my desk at 10:00 a.m. and asked me to join him in his office. He handed me a Payroll memo that confirmed a salary increase of fifteen percent, and a job level increase of two grades. I went from Designer-5 to Engineering Designer-7. My jaw hung open.

Howard chuckled. "I wanted you to see this before checks are distributed today, and to apologize for how late it is. We've all been busy, sometimes too busy, but I wanted to give you the news myself rather than just let you see a different amount in your check."

"I don't know what to say, Howard. Thanks doesn't sound like enough."

"Don't thank me for delivering what you've earned for yourself. Just keep on doing the same."

The full impact of his actions didn't hit me until later that morning when I saw my check. My fifteen-percent raise was retroactive four months, and applied to regular hours plus overtime. It was beyond anything I had expected. Santa wasn't the only one who made that Christmas memorable. Another occasion I'm not likely to forget is the day in January when I met my ex-supervisor in the hallway. It was the first time I'd seen him since the move to Advanced Engineering. He had a snide sound in his voice as he inquired, "Are you enjoying your new job as much as expected?" I told him all right—two levels, the pay increase, and retroactive. That time it was <u>his</u> jaw that touched the floor.

I've heard people bad-mouth Howard. They thought he was too hard on people, or difficult to deal with. I admit that he wasn't always easy to satisfy, and concede that I never wanted to disappoint him.

Howard pushed hard, and expected a lot, but he was always straight with me.

I obtained permission, from the landlord of our rent house in Drexel Hill, to paint the interior. He'd buy the paint; I'd provide the labor. As I've often found, I should have made him put something in writing; two months later he listed the house for sale. We had to buy it or leave, but his price was high and he wouldn't negotiate. No one at the office knew when ground would be broken in King of Prussia, but it didn't seem likely that I would be working there for years. Homes on the New Jersey side of the Delaware River were more affordable, Levitt and Sons, of Long Island fame, were developing the area. And access to Drexel would be easy from there. We bought a three bedroom colonial house in Willingboro, New Jersey.

When I first moved there, a farmhouse that had belonged to Ben Franklin's son was still standing. It fell to the bulldozers three months later. That was Levitt's third major development and the town underwent a name change to Levittown, NJ. It lasted for three years, until there was enough voting residents to change it back to Willingboro. Levitt didn't get mad—he got even. He'd always said that he'd turn his offices over to the town as a Community Center when they were finished building. But when the people opposed him, he reneged and sold it as a commercial office.

In Advanced Engineering I worked on numerous design studies whose results I've seen and followed over the years. The Rogallo wing is one of them. An easy way to visualize its shape is to take a cone configuration, split it down the long axis and rotate the half sections outward 180 degrees until the shape resembles the hang-glider of today. A Rogallo shaped parachute was considered for use in landing manned and unmanned space capsules. Its weight lifting capacity was excellent, and any reentry vehicle recovery would be simplified if the chute could be made steerable.

As well as consideration for parachute recoveries, we studied it as a potential method for launch and recovery of a cone shaped capsule.

The cone would be launched whole; for reentry, half of the cone could serve as a manned return vehicle—sort of a space sled. Half cone shapes were referred to as 'lifting bodies,' because of the lift generated by making the bottom surface flat or nearly flat. A pure half cone shape was unmanageable; unless fins, flaps, or gas jets, were used for attitude control, the vehicle could not deviate from a ballistic path.

I proposed the study of an alternate form of control flaps. It came to be known at GE as a 'palf,' flap spelled backward, because it was the reverse action of a flap. Instead of extending a flap into the super hot airstream, pulling a control surface inward provides a similar control surface effect. Severe temperature exposure erodes the heat protection materials on a flap; a palf would see less heat—less erosion. But, when no development contract came from our studies and presentations, work on Rogallo and lifting bodies was dropped. The X-38 planned by NASA as an emergency recovery vehicle for the International Space Station bears similarity to a half cone, but the side edges are curled up for the elevator effect needed to improve stability.

Another of my assigned projects was to design a chimpanzee couch with arm restraints, so their hands could not reach urine and feces collection containers. The animals were trained to respond to lights, and hit switches, so the observers could determine their ability to function. Monitored by camera, the chimps soon learned that they could get more attention and food from the human attendants when they smeared feces over the mirrors or camera lens. Twenty of the animals were housed in a trailer behind the main building. Jocko, a large male chimpanzee, would masturbate whenever he smelled a female in the area—animal or human. And, believe me, his sense of scent was highly developed. One of the technician-handlers used to go out of his way to show him to unsuspecting ladies. In time, the animals were presented to various zoos. Jocko took the longest time to be placed. He had to be sent where the animals lived behind glass, not in an open-air environment.

With my experience in the Arming and Fusing section, I was asked to collaborate with Tim Martin on a proposal that included rocket-sled tests. After the system had been designed, He was working up the specifications and cost for the test runs, and had calculated that twenty-four sled runs were needed for testing of the concepts presented. Each unit would mount in different angular positions, to simulate different paths, and was hurled down the track into a concrete block target.

Tim didn't realize what a can of worms he was about to open when he asked me, "Check these figures. Tell me whether we need one concrete block, or two, to stop a nose cone of this much mass. I don't think one is safe."

The concrete target blocks referred to were cubes of concrete with metal lifting lugs cast in place. It took a crane to hoist them into position. I ran through his calculations, and said, "You're right. One is marginal. Better use two." His input, with my backup, carried us to the office of the Department General Manager, Hilliard Paige. We were called into his staff meeting and asked to justify the cost of forty-eight blocks instead of twenty-four.

"We've never had to use two blocks before," one manager stated bluntly.

"It adds twenty-four thousand dollars to the bid," declared another. "A thousand for each block, because of the crane and extra handling."

Mr. Paige held his hand up, and silence reigned. "The proposal has to be to the printers by tomorrow," he said. "All we want to know is—are you sure of your numbers. Do we need to use two blocks?"

Tim said, "The calculations are marginal. I think we ought to play it safe. Eric checked my figures and he feels the same." He looked in my direction, to share the pressure he was feeling.

My mother's pet expression, —in case—came to my mind. I clutched at any idea, and said, "Why not show the extra block as a contingency cost, in case the second block is needed for range safety. After the first run, we'll have proof of whether the extra blocks are needed."

"That sounds reasonable," agreed Mr. Paige. "It won't show on the bid price except as a contingency cost." Once he said that, others chimed in their agreement.

Back in our office we were razzed for being brought on the carpet, but when films of the first run became available Ted and I were vindicated. The tip of the missile tore its way through the first block, nudged the second, and flakes of concrete flew like hail in a storm. The second block wasn't damaged severely, and a different face could be used as the second block for successive tests. Over the length of the contract, only three extra blocks were required.

Once again I'd underestimated Murphy's Law. In spring of 1959 construction started on the company's new building at King of Prussia, and six months later I was traveling forty-eight miles each way from my home in New Jersey. GE's Space Center is adjacent to an intersection of Pennsylvania Turnpike and the Schuylkill Expressway, which leads to Philadelphia. Local residents refer to that Expressway as the world's longest parking lot. Three GE buildings are on top of the hill, and King of Prussia Mall is at the bottom. The road in between is named Goddard Boulevard in memory of Robert Goddard, the first American rocket pioneer. In recent years, another shopping center has been built east of KP mall, and joined to it by second and third story enclosed walkways. On my last return visit, it took me a half-hour to find the boulevard I once knew so well.

Outer space was still largely unexplored in the early 1960's, but the drama and tension continued. December 19, 1960 a Redstone rocket with an empty Mercury capsule was launched along a ballistic arc. The second flight contained a three-year-old chimpanzee, named Ham. On April 12, 1961, a Soviet cosmonaut named Yuri Gagarin became the first human hurled into space by a ground-fired rocket. He orbited the earth and was weightless for ninety minutes. Alan B. Shepherd became the first U.S. astronaut on May 5, 1961. Twenty-three days after Gagarin's flight, Shepherd was launched in a ballistic arc by a Redstone-3. He went 116 miles high, did not enter orbit, and was weightless for five minutes. However, with that manned Mercury success, President John F. Kennedy threw down the gauntlet. In a special message to Congress on urgent national needs, May 25, 1961, he said:

> "I believe that this nation should commit itself to achieving the goal, before this decade is out, of landing a man on the moon and returning him safely to earth. No single space project in this period will be more impressive to mankind or more important for the long-range exploration of space: and none will be so difficult or expensive to accomplish. We propose to accelerate the development of the appropriate lunar spacecraft...develop alternate liquid and solid fuel boosters...funds for other engine development and unmanned explorations—explorations which are important for one purpose which this nation will never overlook: survival of the man who first makes this daring flight. But in a very real sense, it will not be one man going to the moon—if we make this judgement affirmatively, it will be an entire nation. For all of us must work to put him there."

The country was being asked to make that 240,000-mile journey in nine years; Congress supported the goal with funds. An eventual 400,000 workers rose to the challenge, and a future was assured for the Gemini and Apollo programs to follow Mercury. In July 1961, Gus Grissom flew the same type of ballistic arc as Shepherd. Upon landing, the main hatch blew too soon. The Freedom-7 reentry capsule was lost in a three mile deep part of the ocean. It was recovered four decades later so the cause for the premature hatch release could be verified. That failure nearly drowned the astronaut. Before the next launch a flotation collar was designed and ready for use.

On August 13, 1961, Berlin residents woke to the sound of trucks and workers arriving to erect the beginnings of what would become a wall of containment. Two thousand workers strung barbed wire. Later the wire was reinforced by cinder block that stretched for one hundred thirty-one miles and survived until 1989.

In February of 1962, John Glenn Jr. was the first Mercury astronaut to achieve orbit for three turns around the earth. Mission Control had him retain a retro rocket longer than normal because they had a caution light indicating that his capsule heat shield might be loose. Scott Carpenter duplicated Glenn's flight in May but his capsule came down two hundred and fifty miles from the landing zone. It was a scary two hours for all concerned.

In 1962, for a speech at Rice University, the President again spoke of the Moon program when he said:

> "We choose to go to the moon in this decade, and do other things, not because they are easy, but because they are hard, because that goal will serve to organize and measure the best of our energies and skills, because that challenge is one that we are willing to accept, one we are unwilling to postpone, and one which we intend to win."

Test pilots are inherently competitive, but astronauts are the ultimate gamers. They race sport cars on sandy beaches or play any game their quick minds can create. It was easy to locate turtles in Florida. Whether they were staying there or not, the astronauts would descend on any motel with a pool. Being the celebrities they were, it was inevitable that when their races were conducted an audience would appear. That was the origin of the Turtle Club. It didn't take long before some wit developed a password with an appropriate response.

When two members met, one would challenge the other by asking, "Are you a turtle today?"

The proper reply is, "You bet your sweet ass I am."

Failure to respond would cost the offender a round of drinks. The game took on added spice when one caught another in the company of prim and proper folk. It became a particular favorite of Wally Schirra. He was a practical joker, an apparent master of the 'gotcha' game.

Walter M. Schirra launched into space on October 3, 1962, but three minutes and forty-one seconds into his flight, Deke Slayton asked, "Are you a turtle today?" Unfazed, Schirra said, "Going to VOX recorder only." He spoke into the tape recorder whose microphone was not connected to an open radio line that the listening audience could hear. His poise stood him in good stead that day; the mission was plagued by small equipment failures.

The sixth and last Mercury astronaut to be launched was Gordon Cooper Jr., on May 15, 1963. His orbital flight lasted one full day. The automatic control system failed, and he had to land the spacecraft on his own. Without a trained pilot at the controls the mission would have failed, and Cooper would have died. After that flight the astronauts were able to convince NASA that man had a better role in space than to be a passenger. They were assigned a larger role in capsule control and the term 'chimp mode,' which referred to being on automatic control, was dropped from usage.

Donald K. (Deke) Slayton, the seventh Mercury astronaut, was unable to fly for years, grounded by discovery of a heart murmur. He continued to fill an important role in developing programs and training the next generation of astronauts.

Another early project for me in the Advanced Engineering office was to work with our computer analysis group. They would run calculations for an optimum reentry shape, but when designers tried to package components, the predicted aerodynamic balance could not be achieved. In reviewing their data, I found the programs too theoretical; they weren't making practical decisions in the selection of materials available in the real world. If a computer said that the design required a metal skin of 0.004 inches thick, or four one thousandths of an inch, it was necessary to realize that was the thickness of aluminum foil—it would not serve well for flight hardware. The smaller the cone-angle, the larger the layout error became. Programmers were then provided with tables of the American sheet metal gages available in industry; when a theoretical thickness was calculated, the computer would substitute the next standard sheet metal gage. From that time forward, our computer predicted designs became more accurate.

No matter how vast a project NASA had been given, to achieve JFK's moon target, there were ideas, tools, and equipment available that could be used as a starting point. Engineering development has been done the same way since the pyramids. Build a model. Test it. Make it better or larger. Do it again, and again.

Nimbus was the only weather satellite that I ever worked on. Our team had to decide how much equipment was required for the task, and how large an array of solar energy cells would keep the satellite operating through alternating periods of light and dark. A tripod support between the package ring that had to face earth, and the solar panels that face the Sun, allow each section to move and rotate independent of the other. That way, sensors on the satellite can keep visual sight of their targets without blocking solar panels from the sunlight that recharges the power cells, or batteries. We won the proposal and another group was formed to complete the design, details and construction.

Other studies included vehicles that would reach Earth's nearest neighbors: Mars and Saturn. At first, orbiter satellites were designed to circle a heavenly body, photograph its terrain, and transmit landscape pictures back to earth. Later ones would land for a closer look at the air and surface composition. Those lander types had to include a means to cushion the impact: use of a reverse thrust engine, or packaging the exterior with a frangible material such as foam, or honeycomb structures of cardboard, which crush at a controlled rate. They had to survive ground impact and be able to perform tests. Internal crushers pulverize surface material, analyze the ingredients, and store the data until a transmission could be made to Earth. Viking was the first vehicle proposed to land on Mars. Although GE didn't win with our proposal, a vehicle did settle on the Martian surface to transmit data on the chemical composition, and thousands of pictures that gave our JPL scientists their first map of Mars.

Advanced types of planetary travelers carry rovers, which can be sent to sample soil and rock composition at locations other than the immediate landing site. Separation mechanisms were developed, as well as decontamination techniques for any part of the hardware that might come in contact with another planetary body. NASA was extremely cautious about any possible contamination source.

We built models of rovers to determine the best way to control movement. In the real world, time before the vehicle responds is critical. Time response for a moon rover is 2.3 seconds for each direction of communication. Other planets being farther away have proportionately longer delay times. Our office team members loved to play with the rover models. We had a tabletop scene built and the rover was equipped to have an appropriate time delay. Many of us stayed after work just to try our hand—over and over. When you saw the vehicle headed for a cliff, a crater, or straight into a mountain face it was necessary to anticipate the delay in responding. It was a good thing the rover model was well constructed—failure occurred more often than not. But it showed that we needed to develop other types of sturdier and intelligent vehicles: rugged, so they could withstand shocks; smarter, to sense danger when a track or feeler was disturbed. We'd try anything except direct human guidance. That didn't work unless the vehicle was moving infinitesimally slow.

I directed construction of the scale or full-sized vehicle models that we designed. Others in the Advanced Engineering unit used them in making solicited and unsolicited presentations or proposals. Military groups were presented with replicas of missiles, and nose cones or reentry vehicles, either single or in multiple clusters—for public relations.

To develop another method of supporting components in the correct relative positions, I worked with the company chemists to develop a foam-in-place technique that used lightweight, rigid polyurea to fill the reentry vehicle's interior. A full-sized model was built to prove the method. It was left open to view by replacing the metal on one side of the nose cone model with Plexiglas. The concept was accepted as a viable method to mount certain hardware.

NASA and science agencies were shown models of vehicles along their interest lines—at first unmanned types that would collect particle samples or expose various eggs or embryos to space environment such as weightlessness and radiation. Animal occupied vehicles followed, and eventually—manned capsules.

A working model that I built for a proposed Saturn capsule and lander became an embarrassment to the project manager who made the presentation at NASA. It was one foot in diameter. With many functions to perform, I used AC motors for the parts that were activated. To replicate the landing, I'd covered the whole capsule in foam. The demonstrator was to drop the model on the floor, let it roll around for a while and then plug in the electrical cord to show what it would do once it 'landed.' Leveling legs extended to stabilize the craft's position, the top of the capsule opened, and whip antennas popped up. I thought it was a winner.

When the manager who made the presentation came back from NASA, he threw the unit on my desk and said, "We sure screwed up on that one."

I was stunned. "What happened? Didn't it work?"

"Oh," he replied, "it worked all right. At the end of my presentation, the program director asked whether GE wanted the contract for the lander, or the electrical cord. My audience cracked up. That was the end of any serious discussion."

I scoured catalogs for DC motors small enough to fit, and large enough to provide the torque required; batteries also had to be fitted into the shell. A Black and Decker cordless drill was disassembled to provide a motor that activated cams for the legs to steady the unit. The interior was rearranged to fit four rechargeable batteries that could be replaced every four hours. The kit included eight spare batteries, plus two recharger units, and our previously embarrassed presenter was pleased with his next reception.

No one had done many of these things before. That's why we studied before anything was built—to develop logical ideas, create hardware, subject samples to environmental testing and measure the results. Like Thomas Edison, we kept trying until something worked.

My spouse met a man living in Willingboro who had his own company to produce TV commercials and short films. He had a GE apprentice that worked for him, and told my spouse, "Sight unseen, your husband can work for me any time. GE trains apprentices to be creative and independent." That was flattering, but I never followed up on the offer. Later I heard that Walt Disney's animatronics unit bought out that company.

Drexel classes were attended three nights each week, at least when I wasn't traveling. Twice it was necessary to drop a semester. Class nights I drove twenty-six miles to Philadelphia, and as much again in order to reach home afterward. Whenever possible, I carpooled to work and minimized wear and tear on the family car. It also lowered the cost for gas or tolls.

Pay was good in Advanced Engineering, and I'd been promoted again. But work required time away from home and family for long hours, or days. Then there were the classroom hours and homework required for Drexel. I met a coworker, Brian Edwards, who had a similar problem of getting his schools together so the company would recognize the credentials. He told me, "GE considers three things to work for them as an engineer: a baccalaureate degree, a state license as a Registered Professional Engineer, or experience as an engineer with other employers. The license procedure in Pennsylvania has a grandfather clause for non-degreed applicants who can prove twelve years of experience."

Brian mentioned it to me because GE had announced a refresher class for those who wanted to take the state exams. Misery loves company. He felt we could help each other, even just to keep our spirits up. We signed up, and each of us wrote an expanded resume that included work and scholastic data. It was to be sent to five references for verification of the twelve years work experience.

Brian and I both passed the Engineer-In-Training exam in 1962 and the practical applications exam the following year. I received my engineering license in mid-1963, but Brian left GE before taking the second exam. Our paths divided. I never heard more about his success or failure. Near that time, my second child, Eric III, was born. I continued to attend class at Drexel now and then, but not all the time. That lessened the expense, and my time away from home.

Project Gemini began in 1963, that design held two astronauts in a ten-foot diameter capsule. It was named for the Zodiac twins of astrology. Grissom and Young flew the first of ten capsules on March 23, 1965. In December, a Gemini-IV astronaut performed the first spacewalk. He had no difficulty getting out, but upon reentry his hose-tether wrapped around him and complicated his getting reseated. It could have become a fatal problem.

We did another study on how to cover the skin of a large vehicle in a cost-effective manner; areas could be covered in different materials. During reentry the skin would protect the craft by ablating or burning away. Eroded sections or tiles would be replaced later, to make the vehicle reusable. I designed other protective concepts: re-radiative metals, carbon fiber mats, and combinations of the same. Someone in the group suggested a Velcro like concept—hooks and loops—except in a heat resistant material. A stainless steel sample was made up. As individual pieces the two halves were flexible, but once placed together we couldn't separate them again. It was just one more thing that wouldn't work. Initially the project was named Space Ferry, and we worked on sequential studies including one dubbed MOL ferry. When it became a reality it was renamed The Space Shuttle.

As Engineering Designers, Dave Weaver and I shared an office; Jim Smith was located next door. He was the business and financial administrator for Advanced Engineering; and often referred to as a paper-shuffler, or bean counter. He had a habit of dropping in to shoot the breeze, and was a champion exaggerator. "Can You Top This," was a comedy show on radio, and Jim was as good at stretching facts as any one of the radio cast. No matter what you said you were doing, he could do it better, faster, or cheaper. If you read a book over the weekend, Jim read two. You're tuning a piano—he was rebuilding an organ. We had great sport in making him reach and stretch for a goal, but one-upmanship was his forte.

Jim was moving on to a new job. We wanted to provide a personalized gift for the occasion. Dave and I were interested in heraldry and created a specially designed family crest: a shield, with crossed shovels on it. Around the perimeter were bastardized Latin words saying, "Seldom right—but never in doubt." Jim either misunderstood what it meant, or had the last laugh. In his acceptance speech he sounded thrilled with our parting gift. He settled into his new position, but I still waited. I had the freedom to recommend studies to be performed—at least when things were slow for the group. But whenever a major proposal came along it was back to the drawing board for me. Salary improved, but my title didn't change.

My forty-eight mile drive to the office in King of Prussia had become tedious after six years. With Sheryl about to enter high school, it seemed a good time to change housing. In the summer of 1966 my family moved from Willingboro, New Jersey, to Kimberton, Pennsylvania. Ten miles west of the office, it was a beautiful drive through Valley Forge Park to reach work. The nearest township to my house was Phoenixville. One of Steve McQueen's early films was a sci-fi thriller, *The Blob*. Filmed on their main street, key scenes took place at a little diner near the end of town. That was where the hero killed the monster, but it was also where I ate with my kids; I have a thing about old-fashioned diners. Shortly after we had settled in to our new house Al Mitnick opened my eyes as to what I needed to do, and Fred Parker gave me my first chance to perform as an engineer. Now I held the position that I'd struggled to achieve, and had to face the problems and responsibilities attached to the title.

◆   ◆   ◆

When I returned for that long Thanksgiving weekend of 1966, I'd brought the children small gifts from the island shops. They tore the wrappings off. Teenaged Sherrie loved the strands of multicolored apple seed necklaces. Rick beat out native rhythms on his steel drum. My spouse and I had agreed that small gift items wouldn't be bought for her each trip—instead, I'd bring a place setting of Royal Doulton china in the Glenhardie pattern; I had brought six so far. The thought flashed through my mind that if December saw an end to the SIV-B program she would at least have a service for eight.

My frequent absence was becoming a cause for dissension at home, and I was regularly reminded that house maintenance and chores weren't getting done while I was having fun in the islands. Fun, hell. The project was stalled. Maybe I was blinded by logic, but I felt that some sacrifice was necessary in order to make a good living for us all. How else could we improve our lifestyle?

Thanksgiving Day was supposed to be festive, but at the moment my mind and career were centered on that structure which was stuck on West Indian's pier. Perhaps it would be better for the family if the program was cancelled, but it was my project. I did not want to fail on any job, and this was my first assignment as an engineer. More to the point, where would I work next if the SIV-B didn't get to Little Buck? No show—no dough—and no job in 1967.

10

*"You only learn to play the game when you play for more than you
can afford to lose."*

—Winston Churchill

GE Vice President Jack S. Parker purchased the Autolite building in
Evandale, Ohio, only to have it sit idle for three years. Often referred
to as "Parker's Folly," the building later came to house GE's aircraft
engine division. When Jack Parker visited King of Prussia, Fred and I
were among those to greet him. Fred was a prankster; no one ever
knew what he would do next. When introduced, he said, "Hi, Uncle
Jack." There was no direct reply, simply a smile and a handshake. It
was a year before any of us knew for sure that he and Jack Parker
weren't related.

Fred called me on Thanksgiving night, shortly after we had arrived
back home from mom and dad's house. He asked if I'd meet with himself and Carl the next morning, at nine a.m. That is, if I didn't have
other weekend plans. When he asked me, I knew there was something
on his mind; I'm not sure when, or if, Fred's brain ever rested. Next
morning, he summarized the project status, and concluded by saying,
"It's imperative that our tests start in early December if we are to
accomplish anything for an end-of-year report. We have ten days, at
most two weeks, to get the SIV-B in the water or the project is finished."

"Sounds like you have something to propose," Carl said.

Fred nodded. "We have two choices. Keep on trying to make our
original plan work, or roll the dice."

Carl leaned forward, listening carefully for each word. "Go on. Let's
hear the one that isn't obvious."

"The crane is still available," Fred started. "Instead of using the barge, lift the structure into the water at dockside, attach flotation gear, and tow the simulator to Little Buck Island."

Neither Carl nor I interrupted. The silence was deafening.

He went on. "It's only five miles. With an open, wire mesh structure like we have, the water drag won't be bad. We can do a really slow tow."

Carl wrinkled his eyebrows. "Is the water at dockside deep enough to cover the structure without touching bottom?"

"Thirty-two feet according to Lars Pederson; the same depth as at Little Buck. The crane can hold the structure in place until we get the flotation gear mounted and balanced.

"You've done some homework," Carl said. "Why didn't you plan to do it this way from the beginning?"

"Because it's riskier. If something should go wrong we could lose our structure. The water depth between our start and finish point goes below six hundred feet. West Indian has insurance on their tug, and they can get a rider on their own policy to cover GE's third party liability concerns, but the simulator itself will be uninsured."

"Hmm," Carl muttered, "I hadn't thought about that."

"Loss of the structure won't help our esteem with management, but insurance on hardware won't buy time. We have to pit or get off the shot."

"In plain English," Carl said, "you're reminding me that if we do this and it sinks the project is dead. But if we don't do this, the time runs out and we're just as dead—it's just something else that kills us.

Fred smiled. "That's it folks. You got the message." He looked in my direction and added, "We have a better than even chance to pull it off. A lot depends on how much flotation gear Eric and I can find, and how soon."

"Why did you invite me here today," asked Carl, "if you knew what had to be done?"

"We're each accountable for the decisions we make in our own area of expertise," Fred said. "I just wanted to know whether we were all in agreement before our necks are stretched out on the block. You know what can happen when they are."

"You're right about the time element," Carl agreed. "The crew will all be doing something else if we don't get to the island soon—at least, those who are still working for GE." He gave Fred a steady look, and then turned to me. "How about your feelings, Eric? Do you think it will work? You haven't said a word so far."

"I was wondering what flotation equipment Camden Marine could locate—fast. And I remember that West Indian had a stock of huge timbers, in back of the buildings at the far end of their dock." Looking at Fred, I added, "Remember when they suggested we might use them as a lifting spreader when the crane first had problems? We could secure them to the structure—like a backbone—to keep it from bending in the middle."

Carl leaned back in his chair, and grinned. "I take it that's a yes vote. Can I help in any way or just leave you guys alone?"

Fred tented his fingers again. "We'll take it from here, Carl. I'll keep you posted. Make a call to Camden, Eric. Let's see what we can find."

They had twelve flotation bags of 500 and 1000 pound capacity on hand, enough for 8000 pounds. But Camden made another recommendation that was invaluable. It reminded me of something dad had mentioned during World War-II, used on what were called Q-boats. To make them nearly unsinkable, the lids and drain plugs of empty fifty-five gallon drums were welded shut. With air trapped inside, they would float. I thought we might also make use of floating barrels; a call to West Indian started their machine shop to welding a dozen drums. I made sketches of the structure, calculated the center-of-gravity, and estimated where the different gear should be attached.

Our crew left for St. Thomas again on Monday. Beams were attached to the top of the structure, and ropes were tied to each end so it could be held parallel to the dock. The crane operator made adjustments in his sling attachment until he could keep the fifty-six foot cylinder level. Thursday morning the unit was lifted over the side. We secured a raft of floating barrels at each end. Pneumatic bags were tied to both sides, at intervals that minimized the beam flexure. The crane cable was loosened occasionally, to check the balance. By Friday, the load was held in a level position by only the flotation devices. Lars Pederson, the Dockmaster, suggested we wait until Monday morning to start the slow tow. A morning start would allow maximum daylight, and an opportunity to refill all of our diving bottles before we left.

Monday morning, we topped off the air content of the lift bags, a towrope was attached, and we boarded the tug. The only thing that made me uneasy about the trip was that the tugboat captain had ordered a crewman to stand by the stern, armed with an axe. He would sever the towrope if our structure became unmanageable. The GE crew was not accustomed to diving in deep water and was ordered not to dive until the tug arrived at Little Buck. Al White and the other professional divers, Dave, Basil, and John, took turns swimming around the structure to watch for problems. They replenished air, as it was lost from the lift bags.

Frank Pugliese made a film documentary. From the hilltop overlooking the West Indian dock, he used a telephoto lens to open with a close-up of the tug and flotation package. The scene was slowly zoomed back until the actual distance was seen. When viewed later, it was breathtaking. To change from a dockside full of structure and tug, until everything appeared to be a straw on the ocean—wow! A small boat ran Frank out to the tug, and he continued to film until the simulator was lowered on the sandy bottom, five hours later.

Shallow waters of the Caribbean are translucent, more green than blue. Underwater vision can exceed one hundred feet. But as the large structure settled onto the bottom of Little Buck cove the water was roiled. Restless grains of sand swirled high above the ocean floor; visibility dropped to half. As I observed the structure in front of me, I realized that in this cove lay the product of eight months love, labor, and devotion on the part of everyone involved. Only a handful was able to witness the birth of Fred's dream.

Jack Burt Entering the Simulator,
After the Tow to Little Buck

Bates Littlehales, an underwater photographer for National Geographic, lived with our crew for a month during this latter part of the project. He had severely cut his right hand on a broken pane of glass; while he recovered, his employers assigned him to what would be an easy shoot. He was with me when the barge availability problem first arose. "When I first started taking pictures underwater," Bates told me, "it was impossible to locate equipment—diving gear or camera housings. Today that's not a problem. Right now, your business is one-tenth of one percent of what they do—not enough dollars for them to care much about. Someday they will change their attitude, but it isn't going to happen today."

His photos appear in one of the four books published annually by National Geographics: *World Beneath the Sea*, James Dugan, 1967. It was unfortunate for documentation and publicity of GE's program that the book's editor, Mr. Dugan, died prior to publication. The material delivered by Mr. Littlehales was reduced to two photos and words of acknowledgement. One picture appears on page 100/101, and shows the simulator soon after it was lowered to the sandy floor at Little Buck Island. The far end of the structure couldn't be seen, but five of our crew are in the foreground. The other picture, on page 201, shows Fred performing anticipated Moonwalk activities. That picture was also the cover for *Oceanology* magazine in the summer of 1968.

Walking the Lunar Path
1966
Fred Parker

Frank Pugliese took similar photographs for the company reports. Fred Parker was in the spacesuit, wearing the Extra Vehicular Activity or EVA Pack in one of its early configurations. His walk took place three years before the first moon landing—before anyone knew that a walking staff wouldn't be required for stability in one-sixth gravity.

Once the SIV-B simulator was in place, emergency procedures were practiced. The crew trained for the roles we'd each assume during an actual situation. It was quite a task to lift a person in a water filled spacesuit without a place to stand and drain it first. I had our land crew build a twelve-foot square safety platform—a timber deck mounted over the sealed flotation barrels we still possessed. There was a shelf built on one side below the water line, for the suited test subjects to use when getting in or out. We could drain the spacesuit before getting the subject completely out of the water.

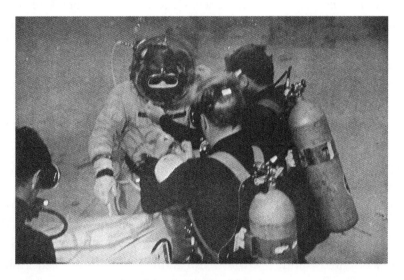

Practicing Emergency Procedures

Pretending to be unconscious Fred was removed from the water to the shelf of the safety platform. Backpack and gloves were removed; he was placed on a stretcher, and into a standby boat. It was practiced until the procedure could be accomplished in five minutes. While any test subject was underwater, a portable oxygen tank was kept at hand on the work platform, and another in the standby boat.

Later, we did have one close call with a test subject; Dick Scoles, a former test pilot, was in the space suit. He'd reached a corner position inside the structure, to pick up a simulated console, when the air supply quit. His back was toward the support diver who was inside the structure with him, but one of the outside divers saw Dick give a hand sign to indicate that his air was cut off. That man moved to the side of the structure, slipped his diving pack over his head, and prepared to "buddy breathe" with Dick.

When the helmet v-band clamp was released, the top of the helmet blew away because of internal pressure, and the outside diver passed his mouthpiece through the wire mesh. The other divers swiftly moved in as they saw the problem and inched Dick out of the open end of the structure. It was over in minutes, but it seemed an eternity. That's what Einstein's' Theory of Relativity explains: time is relative to what is going on at that moment.

For future reference, data was recorded for water drag on a test subject. We determined that when motions were kept below a velocity of one foot per second, drag could be ignored. Movements made by the astronauts are always deliberately slow and cautious; test subjects did the same. We began simple tasks, working in pairs. I was dive partner for Frank Pugliese, and hovered behind him while he filmed the test action. It was necessary to keep from flipping sand as my feet moved, or else it would be seen in the film. The bright yellow paint on the structure took on a green tone in those shallow waters. To my science fiction oriented mind, the wire mesh frame always resembled a birdcage for Sinbad's mythical Roc.

Rather than performing tasks in bits-and-pieces and adding up total time, the large size of the structure permitted whole task or continuous simulation. To prepare the simulated hydrogen tank for shirtsleeve operation by astronauts, a subject was to move through the airlock tunnel into the spent fuel tank, travel the length of the tank, and seal the openings. An anti-vortex screen was to be removed, and two expandable plugs inserted. An astronaut in a pressurized suit, using only ordinary hand tools, was expected to handle those tasks. Once the tank was sealed, the test subject was to travel back into the airlock and through a side hatch to retrieve consoles mounted on the airlock structure. Piece by piece, the hardware was moved through the airlock into the main tank area to set up a laboratory. It was repetitious, and boring. NASA's time-line analysis allowed 126 minutes for the operation. Actual times proved an average of 42 minutes was required, one third of the time predicted.

But there were complications to consider, and report to NASA. As the fuel tank was then designed, the diffuser screen was held in place by sixteen number ten screws; i.e., less than one quarter inch in diameter. With little or no finger sensitivity when working in pressurized gloves, loss of one or more screws would be possible. The loss of even one piece of hardware would condemn the mission to failure.

There was a feeling of being stranded whenever anyone was in the middle of that huge open tank; the test subjects felt intimidated. At first they became helpless when they were away from the sides. Humans are more comfortable with walls or other surroundings that define an area. Without them, they can develop vertigo. We erected a twenty-two foot net across the diameter of the tank, to divide it into two sections. It was accomplished, but our test subject was fatigued from the amount of effort that it took.

Concern that an astronaut would become stranded in the middle of the tank, and unable to reach the perimeter shell, resulted in development of a unique mobility tool. A series of concave and convex beads were strung on a thin wire cable. One cable end was encased in a gun barrel like tube, and the other end had a suction cup. Firing a carbon dioxide capsule propelled the line to the side of the tank. Using a ratchet mechanism the cable pulled those once flexile beads against each other, and it became rigid like a pole. Then a test subject could pull hand over hand to safety. But it was one of the first tools scrapped.

There were routine jobs to be performed in order to keep the project running without a hitch. Everyone had to keep their personal equipment in good condition; the main task was to wash off salt water every night. Best way to handle that was take it with you into the shower at the end of the workday. But air bottles had to be taken to and from the dive shop for overnight filling, plus picking up sandwich makings, filling water bottles, and so on. We all took turns doing those jobs on alternate days, and pooled our money to buy mid-day food items.

The entire crew was staying at Morningstar Beach at that time. Half of us drove into town in the Volkswagens. After breakfast those people would bring something edible for the late sleepers, buy lunch or snack foods, fill water jugs, and pick up the fresh air tanks. They would drive to West Indian dock and load the boat. The back door of every room at Morningstar opened onto the sandy beach. When the late sleepers saw Al's MerCruiser coming around the shoreline they would wade out and climb aboard. Tomorrow it would be their turn to do the reverse. It was safe to leave the cars at West Indian, and everyone including occasional visitors like Carl knew where they would be. Keys were left at the watchman's shack in case someone new arrived.

Sundays were free time now; there wasn't the pressure that we'd felt before. On one occasion, the gang spent the day at St. John's going through the snorkel trail—it's a Federal Park. Art Rachild and I had a laugh, over diving six days a week for work and spending the seventh back in the water at St. John's snorkel trail for recreation.

Dave, Basil, Al White, and Bates Littlehales spent their first free weekend chumming for sharks. Bates had heard tales of a great white being seen in some location and convinced the others to check it out with him. They would throw cut up fish into the water to see what interest they could rouse. They didn't find any great whites, but brought up an eighteen-foot tiger shark that got annoyed when they stopped throwing the chum. He rammed the boat several times; hard enough that Al became concerned whether it would crack the fiberglass hull. Basil had a spear that used a shotgun shell in the head. He killed the shark to stop the ramming. Opened up, that pair of jaws was large enough to slip over Al's waist. Dave and Basil staked the jawbones out on the beach at Morningstar, for the crabs to eat the meat. Someone stole the jaws during the night.

It may sound odd, but steak and lobster can wear on you if they're on the menu every night. I took Eddie Sienko with me and prowled one of the back roads above the waterfront. Sure enough we found a place called King David's Deli, which was open until nine p.m. Once or twice a week we'd head over there, usually when we had the duty to dump the air bottles for refills. Stew, meatloaf, cold cuts, potato salad or slaw—whatever they had—always came with good Jewish pickles. I'd left Brooklyn, but the tastes never left me. I've always been able to find what I want to eat in a Jewish or Italian deli.

The expense of working in a tourist location didn't hit us for a while, but the meals were costly. It hit Fred and I double; anyone in the position of being able to declare overtime could not receive payment for the extra hours. We finally proved that there was an inequity, that we were both unable to recoup meal expenses. The accounting office didn't allow us to exceed the maximum per meal; to compensate, they did allow Fred and I to declare four meals per day on our expense accounts. As I stated earlier, I'll never understand the mental process used by accountants.

Turtle races were run in the pool at the St. Thomas Hilton Inn, where we could always watch 'Wally's turtles.' Thursday nights gathered a crowd. More than a few cash bets, or rounds of drinks, were won or lost there. I saw a few of the NASA personnel attend, but never saw any astronauts there.

All of the hotels, motels, or resorts have big profit motivation; bottled beers, or rum and coke cost a dollar. At the time it only cost seventy-five cents for a whole bottle of rum, and soft drinks were imported. Drinks were served with mostly rum, and a dash of coke. That's a bit stiffer than most people like; two such potent drinks could be a whole night's ration. Our crew limited their intake, and no one tried to keep up with Basil or Dave.

We were introduced us to Frenchtown by Basil; the beer in that section was only twenty-five cents for any brand of bottled beer they had. I like Lowenbrau and I've never bought it anywhere else as cheap as that. As a treat one Sunday night, Basil said he'd take us to a native restaurant we'd like. We drove in three cars, but I know I'd never find the place again. I followed his car. The road got narrow, pavement ran out, and branches swept across the top or sides of the vehicles. At a well-worn building in a small clearing, we heard music playing.

Basil left two parking spaces in front of the building for us, and said, "Go on in. I'll park on the end of the row."

Five North Americans, looking every bit like tourists—we wandered through the front door of a native cabana. Singing and music stopped. Eyes turned, and not all of them looked friendly. It couldn't have been more than a minute before Basil came inside; it just seemed longer. When the proprietor and others spied him, everything returned to the normal bedlam of talking and singing. He introduced us, and it was a long, wonderful evening of food and drink. The best surprise was that our bill for the night was only five dollars apiece.

Apparently our reported findings on the difficulty of closing vents, and separating the tank into chambers, met with agreement from NASA. They made several changes to the interior of the SIV-B tank in order to correct what could have been fatal flaws in the plan to seal an empty fuel tank. At the time we conducted the underwater work our project was officially designated by the working name of The Orbital Workshop. When Skylab was actually flown, the mission was proof of those ideas. Two grids were mounted across the diameter of the fuel tank to create three levels for living, working, and exercising. Self-closing or sealing methods were built into the tank lines. The exterior carried folded solar panels to provide power, and partially shield the exterior from the heat of the sun. But the SIV-B tank contained no fuel for the actual Skylab launch—it was a fully functional laboratory when the flight left the ground.

On the day before the crew left for Christmas, everyone was allowed time for shopping. The variety of name brands was from all over the world. Like kids in a candy store, the gang took off in all directions. Later, we all drove to the airport where I said so long; I had to stay one more day to return rented tools, air bottles, and pay bills. Hanging around with divers and dockhands had affected everyone's language. Many cussed like stevedores, and I reminded them to clean up their language at home.

We were going to be away several weeks, and needed to tie up loose ends for this year. The bill for that one extra night at Morningstar Beach motel shocked me. It was the first night after the winter season price change when every motel, hotel or hotel doubles its room rate until a week after Easter. My chores were finished early in the day. I drove here and there to enjoy the beauty and warmth of the island. Then at one point, I heard a group of children in a schoolyard who were singing Christmas carols. I parked to listen; it seemed strange to hear them singing Jingle Bells to the beat of a steel drum.

A year-end report was written that detailed the establishment of our test station, a resume of crewmembers available for its operation, and conclusions derived from the tasks that had been performed to date. A few days later, Carl told us that funds for 1967 operation were assured.

We'd been working for a month in the underwater structure. It was apparent that the algae resistant recommended paint was simply that: resistant, not algae proof. Starting like fuzz on a peach, thin strands of hair like algae soon became visible, and felt slimy if touched. At first, I'd assigned whoever wasn't otherwise busy to run a sponge over the worst areas. But when the astronaut task schedule began in January, there wouldn't be time to do much cleaning. I needed to address the long-term care procedures for my structure before it became a bigger problem.

11

"Drama is life with the dull parts left out"

—Alfred Hitchcock

Charlotte Amalie is the Capitol City of St. Thomas, the major urban area on that small island. It houses the government offices, airport, post office, hospital, and—of most importance to the local economy—the shopping district. The remainder of the land is rural, with mountains, trees and bushes covering the central part. Year round residents like their seclusion. The hillsides are dotted with homes; many are only accessible from dirt roads. The island shoreline is rocky. Sheltered bays occur around the perimeter, where hotels, motels, golf courses, restaurants, dress shops apothecary, or gift stores cluster to form mini-towns. No matter what else a store offers, they display liquor and cigarettes alongside their fashionable merchandise.

Officially the island is a Territory of the United States, and a governor is appointed by the U.S. to oversee local affairs. By its terms of purchase in 1917 St. Thomas must be maintained as a free port. That is: fewer limits on what constitutes duty free merchandise, and alcohol may be taken home in large quantity. Standard shipping containers hold six one-fifth-gallon bottles. Five slots hold the duty free allowance, but a customs office never collects fees less than one dollar. Any merchant will recommend a bottle of wine to fill the otherwise empty space.

Main streets through the shopping district follow the gentle curve of the shoreline. Streets run parallel to each other, carved as terraces into the rising hillside. Intersecting streets, at right angles to the shoreline and the mountain slope, have deep gutters for water runoff. Seasonal rains arrive in heavy bursts—a violent downpour for twenty to thirty minutes. Those open sewers rush surface water down to the concrete quay that runs along a two-mile length of the shoreline.

When I first arrived, Al White warned me never to stop in an intersection during a rain; he said that it could be dangerous. I took his word for it, but didn't fully understand his concern at the time. On a later occasion I saw a Volkswagen hit broadside by a deluge of water roaring down an open gutter on one of the side streets. The car was shoved across two lanes of traffic and came to rest with its wheels against the curbing on the quay, where the boat tradesmen set up their merchandise in good weather. A few inches of concrete was all that kept the vehicle from being washed into the bay.

Combining vacation time with holidays was common practice for employees of Missile and Space Division. A large number had transferred to King of Prussia from other GE locations, and used free time for family travel over the holidays. The week between Christmas and New Years would offer a long recovery time to our underwater crew. Four people were to be out until the New Year; the rest of the group was at work during a quiet week.

The SIV-B underwater program now had a budget for 1967, work would continue. I started making plans for our long-term presence on St. Thomas and Little Buck. Several radiotelephones were rented—those heavy instruments were the ancestors of cell phones. Although bulky, the phones provided communication locally and to King of Prussia. It was necessary to be brief in our contacts because only one person at a time could talk when a button was depressed, and everyone in the islands who had another radiophone could hear whatever was being said.

Carl assured us that management showed interest in what had been accomplished so far, and our new budget did allow funds for expansion. The two former test pilots could be used to relieve Fred, who until now was the sole test subject, and there was an improvement in safety: an on-site nurse had been added to our original funding request. GE had Industrial Nurses on staff, and I spoke with the Director of Nurses to have one assigned. Catherine Cope was a young widow who wouldn't have a problem traveling for weeks at a time. She became more than a nurse to us; more like a mother hen. Lunch planning was among the many extra duties that Catherine took over as a personal responsibility, and we were glad to let her do it.

Fred told me, "I need to be both a test subject, and available on a minutes notice if Carl can bring or send NASA visitors our way. Arrange housing for the crew, but don't worry about space for guests; they'll take care of that for themselves. I'm going to bring my wife and son with me to St. Thomas for three months."

"I thought your son was in school. Won't that be a problem?"

"Just preschool. He's four, but it will be better to bring my family with me. I was going to suggest that you see if your family would like to do the same?"

"Me?"

"Well, you're second in command. It would be a help to keep you available on the site. You know the hardware better than anyone else."

"What do you mean? Second in command," I stammered. "I'm flattered that you think so, Fred, but no one has made me a manager."

"Maybe not officially, but everyone takes direction from you when I'm not around. Jack, Charlie, Bob—they listen to you because you have more experience, and you're older than they are." He laughed, and added, "Come to think about it, you're older than Carl or myself. You're the oldest of us all. The crew looks to you for guidance. It would help if you could come down with your family, too. At least to get things kicked off."

"Well—I have a daughter in high school, Fred, plus a boy close to your son's age. I'll talk with the family to see what we can work out."

The idea appealed to me, but my spouse opposed everything from the first mention. What I thought would be living a dream; she called being out of my mind. Any suggestion that mom and dad could stay at the house with Sheryl, or that a girlfriend was thrilled at the thought of Sherrie staying with her, met with stern opposition. My wife had never flown before, and didn't intend to start now. She didn't like boats any better, and it's hard to reach an island unless you use one of those means of transportation. Her response wasn't a simple, "No," it was, "Hell, no! And don't bring it up again." I had to tell Fred that I couldn't be in St. Thomas full time; there was an unexpected domestic problem.

"Well," he said, "lets make the best out of it that we can. You're no more than a day's flight away if we have a major equipment problem. Let's plan for you to fly there as needed, but start with a two-week down and two-week home schedule for now. When you're in King of Prussia you can work on any new or modified equipment designs as we determine a need. Work up a duty roster for you and anyone else needed for the test plans we've scheduled. Try to have yourself available at those times. Then, locate some kind of group housing for the average number that is going to be on site at any time. We need to keep housing costs down—living expenses are going to be high, we're in the peak of the season."

Through my island sources, I located two condominiums with four bedrooms each. One would serve for Fred and his family plus management transients like Carl. The second would serve as a dormitory for the majority of transient crewmembers. Motels would take care of any temporary excess GE crew. I also rented a small office on the waterfront—seven hundred square feet. It served as a base for communications, general storage, business meetings or conferences, and we could store special equipment or tools that we didn't want to drag back to the condo every night.

During the week between Christmas and New Years, I gave thought to how that unexpected algae growth could be kept from interfering with the test schedule. Talks with the company chemists, and others at pool companies convinced me that the only way to use a chemical cleaning approach would be to confine it around the structure for twenty-four hours. That led to a design I refer to as the world's largest shower curtain.

I sketched a two-foot diameter by thirty-foot long, tubular float that suspended a thirty-one foot high sheet; it was all constructed of Herculon, a material with mesh reinforcement in the plastic. Two such units were to run lengthwise down each side of the SIV-B structure, and one across each end. That enclosed the smallest area possible. Grommets were provided at three sides to lash the edges of two vertical panels together, and along the bottom edge so it could be staked into the sandy floor. The floats were fluorescent orange colored so no one could miss seeing them. My departure to St Thomas had to be delayed three days, until the panels were placed on order.

Ten days later, my panels and chemicals arrived at the island airport. On Saturday afternoon, after we completed work, the six floats were filled with air from our diving tanks. Towed into position, the panels were joined at the side, seams staked in place, and a chemical cleaner emptied inside the man-made container. Sunday afternoon, I went out on the Mercruiser with Al White and Art Rachild to tow the floats away, and let tidal action clear away remaining chemicals. Monday morning it was necessary to wipe away some die-hard algae strands, but by then it did have eight weeks of growth on the framework. From then on, every two weeks we repeated the procedure. There was almost total algae removal, but minor touchup by hand was always a requirement.

Some tests required more than one astronaut to be submerged at the same time. That didn't occur often, but it required more support. Al White arranged for extra divers on four occasions when two astronaut-suited test subjects were working at the same time.

When I'd first arrived on the island, one of our resident contractors told me, "Don't leave St. Thomas until you've seen a sunset at Shibui." That was a motel composed of Japanese style cottages and gardens. I'd seen signs depicting it on my way from the airport, but never had enough free time to follow his advice.

At the time I'd answered, "We hardly ever finish at Little Buck before dark. I've seen numerous sunsets from there. All of them are beautiful."

"They're pretty all right," he returned, "but nothing like Shibui. That's an experience. Don't miss it."

I'd planned to take his advice, but there never seemed to be time in the evening. Now that the test site was operating six days each week, twelve hours a day, job pressures began to disappear. As testing became more routine it became possible to schedule an occasional day off for each of the crewmembers. That helped lower overtime costs, and rewarded personnel for hours spent setting up the test site. After getting the work crew off in the morning, the others had the use of one or both Volkswagens. But they did have to stay in touch by radiophone, in case plans changed, or there were some emergency.

Art Rachild, I had a scheduled day free. There was no work on site that required a nurse that day, so Catherine Cope joined us. The plan was to visit several beaches around the island and skin-dive on different reefs. At dinner the evening before, I mentioned my desire to see what was different about sunset from the hillside. Art volunteered to go if the drinks were on me. Catherine added that she was game.

When the launch left the dock with our coworkers, we drove around the shoreline to the north side. Art and I dove at four of the beaches, and Catherine tended the radiophone while she acquired a tan. There is a significant difference in the current and high surf condition on the Atlantic Ocean to the north and northeast of the island, than that found on the Caribbean beach exposures to the south and west. We rested in the sun to dry off, and had a late lunch at the nearest hotel. Rather than return to the water, we chose to shop. One of the specialty stores in town is Caron's, owned by the family of actress/dancer Leslie Caron. They carry one of the largest perfume assortments.

When the leisurely day began to wind down, we dropped Catherine at the Morningstar beach motel where she was staying, and proceeded to the condo Art and I were using as home away from home. Cleaning up the equipment and us, we then dressed, and returned to pick up Catherine. Then it was on to Shibui. Driving uphill, and around the switchbacks of that hillside, our Volkswagen struggled to make the grade. The steepness, and the hairpin turns, forced me to shift into first gear. As we crested the hill our destination came into view. The parking area was also at a steep point of the hill, and I placed the car wheels against a rock border.

In the foliage-lined entrance, the odor of native fruits mingled with aromas from the garden flowers; the stronger scents of gardenia and jasmine came through. Our path branched. On the right were cottages with peaked roofs and sliding panel sides making them authentic to the Japanese motif. The left trail led to an A-frame building, and a sign identified it as the refreshment center. On St. Thomas that was an obvious reference to it being a bar.

"Go on out to the observation platform," I told my companions. "I'll bring Pina Coladas." They proceeded across the room, and through a sliding screen wall. Bartenders in the islands like to convince you that you always get your money's worth in their place. They fill glasses to where the liquid is higher than the rim; if it weren't for surface tension the liquid would overflow. The best way to carry a tray full of such drinks is not to look at them, but concentrate your attention on walking. A wooden railing edged the observation platform, and it was fortunate that I reached the railing before I looked at the scenery. There wasn't any ground below the platform for three hundred feet. The building and platform were supported from the rear by struts angled out from the mountainside, and the face of the cliff was behind me.

If someone were to check today, I'm sure they'd find my fingerprints on the railing. I closed my eyes, took several deep breaths, and began to relax. Japanese melodies played in the background and helped to sooth me. I opened my eyes. The panorama included the harbor. Charlotte Amalie bay was like an upside-down moon, with Frenchman's Island lying between the crescent points. Run-down docks edged the near right. They hardly seemed worth the fifty million dollars that the U.S. paid in 1917, but those were the former submarine docks; today they anchor only tugs and barges. Requiring fewer deck-hands than freighters, barges are more common in the inter-island trade.

Harry Truman airport was visible to the far right. Faint sounds of the huge fans could be heard. At the western end of the airport, a sand dredge was pouring fill against the island's edge. At that time, takeoff was uphill, and the runway ended in a two hundred-foot drop. If an aircraft wasn't equipped with a Jet Assisted Takeoff cartridge, or JATO, they couldn't leave with a full tank of fuel; those planes had to fly thirty miles to St. Croix and top off their tanks. In later years, the runway was leveled and extended.

To the near left Yacht Haven was visible, with a collection of masts against the backdrop of West Indian Company sheds. Two small cruise ships were at the dock, with lights blazing on their decks and masts as a beacon calling to the late shoppers due back on board. Anchored in the mouth of the bay was a liner too large to bring up to the pier, and two launches shuttled travelers to their floating hotel. A flat-bottomed power craft followed. Easily identified by candy striped roof supports, it held a precious cargo of tourist purchases when liners were outside the harbor. On other days, it served the island population as a funeral barge for burials at sea.

It was the time of day for passengers to change into dinner dress. The liners remain until full dark. After dinner, when entertainment starts, the vessels slip their moorings, and debark in silence. They travel at night, and the tourists awaken to another day in a different port-of-call. On my extreme left, cable cars ascended the mountain; as sunset drew near, Aton's devotees moved from the beaches to the heights—where gods have traditionally been worshipped. The faithful change emphasis from tanning their bodies to watching their god settle in to rest for twelve hours of the night. At eye level now, the sun continued to plunge.

I watched in awe, as from a cloud. Still holding the railing I sipped the cool refreshment, like one might drink from a chalice. Ocean and sky ran uninterrupted and met in a faint line at the horizon; it was difficult to see where each ended. A twin now mirrored the real sun in a ripple free ocean; the two blazing spheres consumed each other as they merged. With scarcely any movement of air, small, translucent clouds hang low over the fiery scene.

The two globes met, clashed and formed an hourglass, but only for an instant. Shape change continued—oval, then circular. As if a giant eye closed, the brilliance subsided. Muted colors bounced between the water and the clouds. Gentle shades of yellow, orange, or purple spread across the horizon, and gray crept in to replace those tones. Though only minutes had passed, the night sky entered, and stars hove into view. We three stood motionless. I'd been holding my breath; now it came out as part breath and part sigh. It was like stepping down from the clouds. I found out later that Shibui is a word used by the Japanese to convey feelings evoked by an autumn landscape. Landscapes in St. Thomas were undeniably gorgeous, but the memory of that sunset, as seen from Shibui, is unforgettable.

Next day it was our turn to work while others made free, but the weather began to turn bad; waves between St. Thomas and Little Buck were choppy. Riding to work in the MerCruiser had always made Art queasy; he'd always had to use Dramamine to settle his stomach from motion of the boat, though he felt okay to work once we were in the water. Al's boat had to be slowed—to time its speed, and match the wave frequency, or else the hull pounded into every crest. That wasn't good for the boat as a fiberglass keel can crack. It wasn't helping the passengers either. Though not pitching like the boat had done, even the safety raft at Little Buck Island was in motion. Little could be accomplished; the crew was limited to working in or around the equipment building. We made a short day of it.

When the water was calm a day later, we went back. Everything looked okay as we arrived; it wasn't until we were on the raft that anyone noticed—the observation tower was missing. Al had gotten ready first, so he dove. It was only a minute before his head broke the surface again. He called for us to get in the water and take a look. The normally dry observation compartment lay upside-down, on the bottom. Scaffolding was strewn out in a lengthwise pattern, and resembled the skeleton of a beached whale. Two legs on one end section were out of line with the others. The welds had broken.

West Indian straightened the tubular legs for me, but they couldn't weld aluminum. To drill and brace the frame using brackets bolted in place would be difficult; it wouldn't produce as strong a joint either. We had to find someone on the island with the capability to weld aluminum. It was Dave Stith who saved the day. Two U.S. Navy ships anchored outside the harbor late in the afternoon. He contacted the captain of a U.S. Navy cruiser lying outside the harbor. When Dave explained our problem, the captain sent a launch to West Indian's dock and brought Dave, Fred and the broken frame to the ship. When it had been repaired, they were escorted back to shore.

It was eleven o'clock by the time the repair was complete, too late to show our gratitude. Dave asked the captain whether the two seamen who had done the welding could join us the next evening, and leave was approved. We took them to Johnny Johnston's club for dinner and drinks. J.J. used to be the singer on Don McNeil's Breakfast Club, a popular radio show originated in California. Our two helpful seamen had to meet a naval launch near from West Indian at midnight. They didn't arrive much before that time even though J.J.'s club is just around the corner, between West Indian and Yacht Haven.

Two days were spent resurrecting the observation platform and tower. I placed the welded section on the very top where it would see the least weight and wave action on top of it. The worst part was to lift the observation platform bowl. Installed the first time it had been done with the observation compartment empty. Now resting on the bottom, it was full of water. We had to lift it using the flotation bags, and tow it to the vicinity of the safety raft. Then, bail out the water, tow it back to the scaffold, and secure it. I placed more ballast around the base to stabilize it, and added additional cables back to the cove floor—in hopes of preventing a repeat of the disaster. We couldn't always rely on a friendly ship to solve our problems.

12

"Gravity is not just a good idea. It's the law, and it's not subject to repeal.

—Rules of the Air: Author unknown

In the summer of 1967, Fred asked me to find ways that would improve simulation techniques, and said that our next project would be the Manned Orbiting Laboratory. He felt it would enhance the team's image to management if we involved our parent project as soon as possible. Performance difficulty of the crew tasks needed to be evaluated; we'd determine the level of severity that Air Force personnel would be asked to perform on MOL space missions. That branch of the military maintained a separate group of test pilots, including those on the X-plane program.

Though necessary to compensate for water drag by using open or mesh-covered frames on our test equipment, hardware for our next tests would be even more realistic. It would simulate not only physical size and weight of the object being moved, but also the mass and inertia characteristics of each object. Adding a plastic sphere near the center of gravity of the hardware accomplished the first requirement; the sphere held an amount of water equal to the weight of the object and provided the mass to be moved. Baffles or barriers inside the sphere prevented the water from sliding around inside the sphere as the package was moved. That produced the inertial effect of the object's mass; when a test subject changes their direction this function is immediately noticeable. The hardware feels like it has a mind of its own and proceeds in a straight line—like a gyroscope would respond to a directional change.

New MOL simulator structures were designed. A short tunnel started between the two seats in Gemini and allowed the crew to enter the laboratory section. They could also exit through external hatches, to move larger equipment between the units. Hatches and tunnels were thirty inches in diameter, compared to forty-eight for the SIV-B simulator. For a test subject in a pressurized suit it was a snug fit. A slimline version of the rebreather was designed for use in those tight conditions, but many tasks were performed while breathing through a hookah style air hose.

In September my son, Eric III, was enrolled in a Montessori school. At the time, pre-schools were in a minority. One requirement for admission was parental participation in one of the classes—any class other than the one your own child was in. Each week my spouse performed as a classroom aide for the five-year old group, and that teacher was reading to the children from a book by Jacques Costeau. For their activity time the class had been constructing items that pertained to the story.

She mentioned to her class teacher, "Eric is doing something like that in St. Thomas. GE is doing an underwater program of some kind."

The teacher was ecstatic. She asked, "Does he have any diving gear, and could he bring it to show the children? They've only seen pictures."

I called the teacher later that day and mentioned that GE now had a film of astronaut activities in the SIV-B, plus some footage that showed Fred in his suit doing the simulated moonwalk. She arranged for me to talk to her class first, and then the film would be shown to the whole school.

On Friday morning, I took Rick to school, and went to meet my lecture group. I'm not sure what I expected of five-year-old children, but by the time I left I'd learned a lot about their capabilities. They had used paper towel rolls to represent dual air tanks, with a Band-Aid box for an air regulator. A piece of leather or rubber for them to bite on represented the mouthpiece, and it was strung together with cord or plastic strips for the air carrying tubes. Several wore a pair of swimming pool goggles over their eyes. Their boat was an outline on the floor, bordered by wooden blocks. The children demonstrated how to roll off the side rail, backward while holding on to their goggle-facemasks. They were eager—full of excitement, and polite—not just screaming questions. Each waited their turn to talk, though they squirmed in anxiety.

One boy asked, "Why do you have to spit in your facemask? My mother says I should never spit."

"I'm not sure what chemical is in spit that isn't in water," I said "but it cleans the glass better. You can use water, but the glass fogs up again faster. We just do it because we know it works."

"How many tanks do you use in a day?" asked a red headed girl. "And how long do you dive?"

"Usually two tanks a day—sometimes three. Each tank lasts about one hour."

"Do you ever run out of air?" asked one shy child.

"No. When it becomes hard to suck in a breath of air we know we're close to the end of the tank." I pulled a lever on my diving backpack, and said, "Then I flip this handle down. That gives me five more minutes of air to get to the surface."

"How deep have you been?"

"We work at thirty-one feet. But I've been down to ninety feet in a stone quarry in Norristown. That's one place where we can practice in this area."

"Have you seen any sharks?"

"Not in the cove where we're working, but—" I told them about the tiger shark ramming Al's boat.

"Any other big fish?"

"No really big ones, but twice I saw a barracuda in the cove. I'm told that they have a territory around their home; they will attack if it is their home. Otherwise they charge toward you and leave. None of us wore anything shiny because that would attract them, and make them attack."

I sensed disappointment in the group—that the dangers of the deep, which they had expected, had missed our crew somehow.

"One time," I ventured, "three of us were exploring around the rocky edges of Little Buck Island and surprised a four-foot wide leopard manta ray. We swam up from below and didn't see him. He came off a rock shelf and soared over our heads." I ducked my head down, and waved my arms to indicate the ray's flight path over the top of me. "The first sign of him was a shadow zooming past. I'm not sure who scared who the most."

That brought giggles from the group. They were an orderly class. The questions were perceptive for five-year olds, with a level of understanding that I hadn't expected to find. It was necessary to break off questions because the school director was ready to show the film. At the end, there were more questions until parents came to pick up their children at noon.

The equipment and film were home with me for the weekend, so I'd also arranged to show the same things to the youth fellowship at my church. Sunday evening I repeated my presentation, and asked my teen-aged audience for their questions. There were none. "Strange, I thought. "With all of the news coverage on astronauts and space, I thought it would be of interest to teens to know how they train."

Later, one by one, boy or girl, they would sidle up to me and ask the questions they didn't want to ask in front of their peers. More than one prefaced their question with, "I don't want to seem foolish but—" Somewhere between the time of preschool and high school peers, parents, or other adults intimidate children. Communication to those young adults was more difficult for that hour than everything I encountered in four hours with those fresh young minds.

I'd intended to place the smaller Gemini/MOL structure on the smaller sandy patch in the cove, but fate had something else in store. A hurricane passed close to St. Thomas, and Al White was asked to check on things at Little Buck. When he called back, we were advised of problems. Someone had done minor vandalism to the building. They'd broken in and written graffiti, but they left the door open as they moved on. There was hurricane damage to the roof, and parts of the interior walls were soaked. Worse yet, the SIV-B structure had been damaged by the heavy surf action.

Fred was on his way to the airport almost as soon as the phone was down. He said, "Expedite that MOL design, it may be all we have to work with. I'll let you know what I find out." When he called back, it wasn't as bad as he'd first thought, but not far from it.

The observation platform and scaffold was wiped out for good this time. Pieces of the tower were strewn around the cove. The simulator structure had apparently lifted from its sandy bed and twisted. One corner of the cradle, at the rear, was a foot higher than the other side. Fred counted ten ring and stiffener joints broken, also toward the rear of the unit. After viewing the damage to St. Thomas' business and residential areas, he felt that our structure was in reasonable shape for what the whole region had experienced.

We did complete a few studies that fall where it was helpful to still have the larger simulator, but it was obvious that it would take too much time and effort to try to repair it. I had the MOL structure under construction' Fred decided that we should clear the larger sandy patch and use it for the MOL tests. We'd originally agreed with the Navy to dispose of the structure when our program was finished, but they referred us to the Coast Guard who had the actual responsibility for where it could be safely jettisoned. They wanted us to send it to the deep six, that is, drop it where the depth was a hundred fathoms—six hundred feet.

West Indian had stored the original flotation gear for us: the air bags, timbers, and fifty-five gallon drums. Those were again lashed in position. The tug managed to get the SIV-B partway to the recommended site before the structure began to crumple, and the towrope had to be dropped. My simulator sank in one hundred ninety-five feet depth, and was added to the Coast Guard's map of underwater hazards.

I didn't have the opportunity to participate in the short series of island tests to clean up the Orbital Workshop project. It was too important that the Gemini-B/MOL simulator parts were completed, plus the containers and lock-down mechanisms for the next round of tests. I was disappointed. It was much like what happened to me while in Advanced Engineering. When something important was taking place it was necessary for me to work on the design—strapped to a drawing board again.

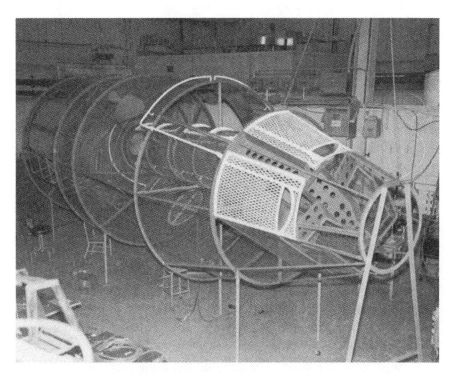

Gemini-B/Manned Orbiting Laboratory, at King of Prussia

When the fabricated sections were shipped for this second project, they were sent on a container ship, from Philadelphia to St, Thomas. The trailer-sized container was positioned at the southernmost tip of West Indian's pier. I left again, with two technicians: Art Rachild, and Eddie Sienko. This time, the assembly would be completed under the water.

With the smaller diameter and length of the MOL structure, it was possible to lower those lightweight pieces from the dock to a small barge. Sections were laid out on the deck in the lengthwise position as to where they would be assembled. West Indian's tug towed the loaded barge to Little Buck, where we lowered each ring section to rest on the sandy bottom. Al White had followed us in his MerCruiser. We donned scuba gear, and the four of us carried waist pouches filled with nuts, bolts, washers, and wrenches.

Like pieces of an Erector Set, we assembled the laboratory and its adapter, into a vertical tower. When the cylindrical unit was assembled, without the Gemini-B capsule, we used ropes to tip the tower over to a horizontal position. Then the Gemini-B sections were lowered, added to the front, and the traverse tunnels were secured between those sections.

Underwater Assembly of MOL
Author, Left
Al White, Right
Ed Sienko, Rear

To assist in neutralizing the weight of the exit hatch over one pilot seat position, the entire mockup was rotated forty-five degrees as viewed from the front. If such details weren't taken into account in the design, it would have been necessary to compensate for gravity in some other way. On this occasion, it took only four days to set up the structure.

Assembly went so well that we were done earlier than anticipated. There was a chance for us to go lobster hunting over the weekend. Unlike the Maine variety, the Caribbean or Longusta lobster has no big claw, but it's just as tasty. Al took us where there were a number of holes to explore at about fifty feet depth. We carried lobster hooks, and dangled them from a thong around a wrist. Reaching in a hole with a hand could be dangerous; you could find a moray eel or sea snake instead of dinner.

We grilled our lobsters for dinner, but Al mentioned an important detail. "Never cut a live lobster—the meat will be tough. Always boil it first, to kill it, no matter how you intend to cook it."

While I was absent those weeks at a time, my wife began a new hobby. She attended Thursday evening auctions that were held in a barn three miles away from our home in Kimberton, Pennsylvania. On auction nights that were near paydays, bidding was lively. In fact, there was a tendency for the crowd to buy at peak or premium prices. On other occasions the bidding was slow and prices lower. Small items were grouped together as the auctioneer wanted to clear out the building. On those nights, a whole box of miscellaneous items might only bring a dollar or two. If you wanted only one item in the box, it was necessary to buy everything.

She developed a system of using slow nights to buy, and on nights near payday she would sell. Arriving early to review the boxes for items, she knew which to bid on. If prices went high—back off. Let someone else have it, there was always the next week. Before long, cars couldn't park in our garage. Removing any desired item from a purchased box, she'd repackage the remainder—always being sure that there was at least one item per box that would draw attention. On potentially good nights she would transport those boxes back to the auction. Her investment doubled and tripled.

Dealers who came for merchandise to fill their second-hand or antique stores began to recognize her. They'd refrain from increasing a bid so she could win. After the auction they'd check with her to see if she would sell or trade something in the box that was of value to them. Sometimes she did, and they both won; it kept prices lower for all. They appreciated that. Others were welcome into the game if they weren't greedy.

It wasn't always boxes of merchandise. She practically stole a Bolen 7.5 horsepower tractor with full grass mowing and snow plowing accessories for seventy-five dollars. In running condition, it was too large for me to transport home by any means other than driving along the edge of the road for three miles. The auctioneer let her store it there for a week until I was back in town. That mower worked for all of the years we lived in Kimberton, and stayed with the next owner. There was a wooden fort acquired for Rick at ten dollars; taken home in panels, it was a great addition to his play yard. Being a teenager now, Sheryl went with her mother occasionally. She located a hopechest for herself, and items to fill it.

In October my spouse outdid herself. Three refrigerator-sized boxes were sold to her for ten dollars a box. Preparing their toy department for Christmas sales, Strawbridge and Clothier's department store in Philadelphia had cleaned out broken items, and those with damaged cartons. Each container was too heavy, and bulky, to fit any vehicle we owned. It took three trips in a borrowed pickup truck to get everything to our garage. There were several identical items, and piles were sorted on the floor. We kept toys that would be Christmas presents for our two children, and the remainder was repacked in smaller containers. At auctions during November and December, the boxes that were resold from that thirty-dollar purchase became two hundred and eighty dollars.

She became knowledgeable about quality merchandise, and developed a good eye that could spot real antiques from the imitations. Offered employment or a partnership by two ladies who operated a local antique store, she refused. They wanted her to handle their retail store so they could be on the road more often and find new stock. She said thanks, but no thanks. To her it was fun, all a game.

There was no immediate third project in front of us in the early part of 1968, but the company never let manpower remain idle. Several team members, including myself, were reassigned to other projects for the Manned Orbiting Laboratory. I moved to another office, and became part of a six-man unit building special effects for the flight simulator that would be used in training Air Force crews. NASA had a spotlight on its missions, but the military didn't use trumpets to declare their intent to continue a role in space. My manager was now Al Little, but all of the team kept in contact with Fred, just in case something changed. All work under MOL was good, but some things were better. None of our underwater group wanted to be unavailable if another opportunity broke.

In November, my old friend and former manager, Al Mitnick, called and asked, "Can you make it to a meeting in my office this afternoon?"

"Sure thing. Can you tell me what's up?"

"There's renewed interest in a project you once worked on. We need to determine the feasibility of further development of an idea, and might be able to do that by using underwater simulation techniques. The brass wants to know whether it is worth spending any more time or money on it. It's just a possibility, but they need general information on the underwater life support pack and the St. Thomas facility." He didn't name the project over the phone; we'd all learned not to say much on open lines. Any discussion about projects stayed in the building.

"Do you want me to invite Fred? He's in town this week."

"No—not yet. This is an exploratory meeting. It can lead to more, or nothing. Come over at two o'clock."

He'd piqued my interest, and for the next four hours my mind ran over the list of my former projects. Which one did they want to revive? I wasn't surprised to see Stan Lipson, the present Manager of Advanced Engineering, or Hal Bloom, another unit manager parallel to Al Mitnick. Then I caught the eye of a smiling consultant by the name of John Quillinan. The light dawned for me; John was the clue. He must have asked management to resurrect the MOOSE project, an acronym for Man Out Of Space Easily.

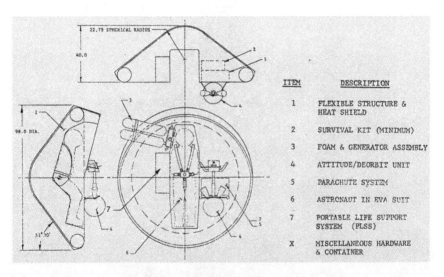

ITEM	DESCRIPTION
1	FLEXIBLE STRUCTURE & HEAT SHIELD
2	SURVIVAL KIT (MINIMUM)
3	FOAM & GENERATOR ASSEMBLY
4	ATTITUDE/DEORBIT UNIT
5	PARACHUTE SYSTEM
6	ASTRONAUT IN EVA SUIT
7	PORTABLE LIFE SUPPORT SYSTEM (PLSS)
X	MISCELLANEOUS HARDWARE & CONTAINER

Manned Orbital Operations Safety Equipment. MOOSE Design

It was one of his wildest ideas. He was full of dreams and schemes, but that one was a favorite of his. I'd documented several projects for him, and worked up material to use when MOOSE was presented at a scientific congress three years back. So far it remained only a possibility.

Stan pointed out that, the more orbital flights taking place, and the closer we come to moon flights, the greater the interest by NASA on safety aspects of the manned space program. I was correct in my surmise that it was MOOSE, but although the acronym remained they'd since refined the project name to Manned Orbital Operations Safety Equipment. That sounded more professional to corporate ears, and hopefully to NASA.

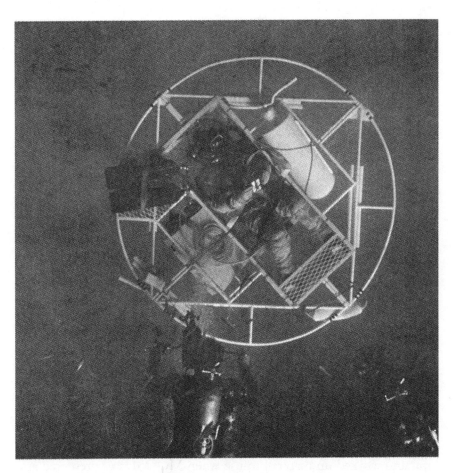

Rigid Framework Simulator
MOOSE

John's concept came in two versions: a storable, inflatable package that could be donned when or if needed—sort of an individual space parachute with a retro-rocket. There was also a rigid model, large enough for one or two occupants. NASA had not orbited anything larger than a two man Gemini capsule at the time, so a one or two person capsule was still a reasonable consideration.

I'd brought photos of the Buck Island facility, Fred's backpack and other support equipment, plus copies of the reports on the two completed programs. John showed me the latest data and drawings for the MOOSE. Within two hours we'd agreed that simulation was possible for the two versions—both inflatable and rigid.

A/E was now a part of Reentry Systems Department. Back in the MOL area I went to see Fred right away, and told him of the meeting. He realized that we needed to keep their interest piqued, and suggested to me that I work up a list of the equipment required. When he had that, Fred would work up test agenda and cost estimates. To help obtain funding, Carl Cording was brought into the loop immediately. Fred negotiated with Al Little to make my services available. I didn't realize at first that part of Al's agreement to my participation created a position for him on the island team. There was always use for another set of hands, and with A/E footing the bill that condition was acceptable to Fred. We were back in the underwater business.

All we were missing was the two test structures. Sharp, pointed, streamlined vehicles reenter too fast for a human's safety. MOOSE had been designed as a sphere-cone shape with a large radius and wide cone angle, to keep its reentry velocity low. In the earlier days of missiles many experimental packages had been flown with that blunt, stable profile—we had reams of data on its reliability.

Inflatable Simulator
MOOSE

The flexible and rigid structures were simple designs. A rigid unit was completely manufactured by Rocky's lab in six weeks. An inflatable rubber model needed more expertise for assembly, both in material and bonding. A West Virginia company, which specialized in life jackets and rubber rafts, built that one. Everything was in readiness by April 1968, but a few of the island support personnel, master diver Dave Stith for one, needed a couple of weeks to clear away other work. The test program was scheduled for three weeks in mid-May of 1968. Work conditions were great; water temperature was in the low eighties—it couldn't have been a better assignment.

The only trouble we experienced was in adding ballast to the test structures to make them neutrally buoyant. The rigid structure was below the water surface, and we needed to place chunks of foam at strategic locations around the diameter. Swimming down to the structure while carrying a buoyant foam block was more difficult than anyone had considered. Only the strongest swimmers could tuck a full block of foam in their wet suit and dive below ten or fifteen feet. It became a game, and most of us were laughing heartily. We broke the foam into smaller chunks and finally managed to balance the structure's weight.

At the end of each day we gathered information from the crew and test subject, in debriefings. That would provide information for Advanced Engineering staff to plan future improvements of both the rigid and flexible type designs. Despite our efforts in working the problem of manned space-flight safety, no further work was scheduled for MOOSE. To-date, no individual emergency device has ever been provided on a NASA launch.

13

*"For all of your days be prepared, and meet them ever alike. When
you are the anvil, bear—when you are the hammer, strike."*

—Edwin Markham, 1852–1950

A Saturn-V with an Apollo capsule stands three hundred sixty-four feet tall. The first and second stage motors launch the module; its third stage and service module enter earth orbit, to fire the spacecraft on a path that intercepts the moon. Grissom, White, and Chaffee were named as the crew, and the first mission was scheduled for February 21, 1967. In practice, on January 27, 1967, an electrical spark caused a fireball that killed all three. It was a tragedy, and a major setback until the cause could be determined. The next manned flight did not take place until December 21, 1968, when the Apollo-8 capsule was launched. The astronauts achieved moon orbit December 24, but it was an observation and photo mission. There was no attempt to land.

By January of 1968, at least three GE divisions had something to do with underwater hardware; so far, ours was the only ones to actually get wet. One unit designed and built a huge buoy to measure wave height and tidal effects. It was anchored off the U.S. coastline by a ship belonging to Scripps Maritime Institute. GE personnel were there to see it launched, but they were only observers.

Another group studied a position control system for the DSRV, or Deep Submergence Rescue Vehicle. That project involved setting up a Quonset hut to simulate the upper surface of a submarine, and using a small blimp to represent the rescue vehicle. Their job was to determine how to maneuver a rescue craft over an emergency hatch on the dummy sub, and secure to it. The program was successful and aided Navy's development program of an actual rescue craft. It affected installation of emergency equipment on submarines, and affected the hatch design for future subs, but no one got near water.

With no program scheduled to run in fall 1968, Fred and Carl talked to anyone whom they felt could use the services of an underwater team. Six months earlier they had met with staff-level people in the oil and gas drilling business. Nothing had come of it at the time, but Fred was now invited to attend a meeting with four of their corporate officers. Through rumors, Fred heard that the GE rebreather unit had been tested by the U.S. Navy, and held an unofficial—spell that classified—depth record. Oil companies were now working in deep-water and that severely limited a diver's time on the bottom.

Rebreather Backpack
Navy Design

Every thirty-two feet of water causes an increase of one 'g' or one times the force of gravity, causing compression of the gases present in a human body. Bubbles tend to have trouble entering smaller blood vessels in the arms, hands or legs. They accumulate at body joints; that's why decompression sickness is called the bends.

The meeting Fred attended was held on a drilling rig in the Gulf of Mexico. They asked him to come several days earlier, so he could be given an underwater tour. He could see their situation for himself, and make informed recommendations as to possible use of our rebreather unit. They employed a mixture of hardhat and scuba divers. Hoses posed problems in maneuvering, and swimmer teams were using triple air tanks to maximize bottom-working time. After a fast ascent, both types of diving rigs required that the diver spend long, inactive stays in a pressure chamber—to decompress the gas bubbles in their bloodstream.

When Fred returned, he told tales of a piping-forest on the seabed. Tree-like valve heads sprouted from drill holes in the floor, surrounded by rows of pipes directing oil to accumulation areas. It was a veritable beehive of productivity that was only seen by only a handful of humans. Drilling industry reps were impressed by the eight hours of working time the rebreather offered, but our unit worked with ordinary air and a small oxygen supply. Their conversation implied that Fred should develop a dual-gas system, that is: air to a specified depth, helium-oxygen mixture below that. Though interested in the benefit from such a unit, they didn't offer to fund its development.

The major idea that Fred brought home with him was that the drilling companies were considering construction of underwater habitats for divers. Once a diver is at a given depth for an interval of time, body gases are fully compressed; decompression takes the same amount of time whether a person has been down one hour or forty. With a dry underwater shelter, workers could rest or sleep between tasks, and return to the surface later. Only one decompression period would be required for workers to remain over extended work intervals.

A year earlier, GE had conducted tests on living and working in a vacuum environment of space. Inside of their huge vacuum chamber, they'd mounted a space station replica, and four volunteers worked under observation for thirty days and nights. It was as close to a space station operation as could be simulated on our one-g planet. Fred let his imagination run rampant. He didn't consider just the needs of oil drilling companies, but marine biologists, the Navy, NASA and other groups who could apply the concept to their own specific needs. That's what he and Carl pitched to our GE management—an underwater test chamber with a small group of volunteers submerged for a long period.

They proposed to sink a chamber in the fifty feet of depth available at the cove of Buck Island. After the first few hours the test subject's bodies would be full enough of compressed gas that long decompression would be needed if they surfaced. It would be a psychological test as was physical; although the surface was close it would be impossible to leave the water without a period of decompression.

The first thing Fred asked of me was to locate the previous structure, judge its condition, and determine whether anything could be done to make it waterproof. Our back lot at the Space Center was a hardware storage yard. The test structure was still there, not far from the high roll-up door adjacent to the vacuum chamber where it had been used. I looked at the station replica, to determine what might be usable. Fred's project was to be called Tektite after the tiny meteors or space particles, which bombard the Earth daily. He asked me to make a few conceptual sketches, and estimate the changes that would be required.

Test panels, internal components, bunks and lockers, were all in good enough condition for reuse, but as a two level chamber it was too tall to stand upright in the shallow bay at Little Buck. The upper section would be too close to the surface. By splitting it in half, the living and working compartments would be in adjacent cylinders that could be connected. If a box-like section were added as a base, it would provide ballast weight to counter lift created by the two air-filled chambers. Racks could be welded on the perimeter, to provide a place for external support equipment. That way, surface divers could check the hardware mounted around the perimeter, and replacement diving bottles could be stored there until needed. Except for radio contact, test participants were to be kept in isolation.

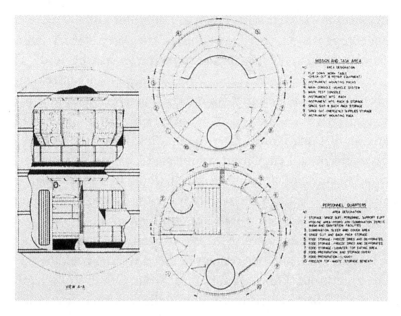

Vacuum Chamber Structure,
Before TEKTITE Conversion

When Fred attempted to acquire the drawings that we needed, to separate and refit the cabin structure, the other manager chose to horn in instead of cooperating. His group was nearing the end of a project. With drawings at his fingertips, and men free to begin work, Fred was forced to collaborate, or design and build everything from scratch. The Tektite program involved use of our rebreather; Charlie Soult, Jack Burt, and Fred began working on the program full time. Carl was involved because of funding. Rocky's technicians built the new hardware, so Art Rachild, Bob Plank and Eddie Sienko were kept busy on their share of the work. With eager participants in the other unit usurping my design position, I was reassigned—again.

This time I worked for Ted Johnson. His primary task was design and construction of the special effects for a full sized cabin simulator of

Manned Orbiting Laboratory. Manual controls, consoles and view-ports were to be built into a realistic training model of the vehicle; power conditioners, slide projectors and other support components would surround the cabin structure. Ted was a great boss. My cowork-ers were the best. Work on that Air Force astronaut trainer was stimu-lating. If I had never been part of the underwater team the MOL cabin simulator would have been ideal; I would have no reason to be upset at working on such a program. But my feelings were hurt. Though my mind understood the circumstances—my heart followed Fred's progress.

The program conducted their tests with five multi-degreed volun-teer subjects who lived underwater for thirty days while they performed regular underwater tasks. It was easy to find out about Tektite's progress, or problems. Meanwhile, I came up with a few underwater ideas that Fred might be able to use. They were offered to him, much as I'd tossed out ideas for studies when I was in the Advanced Engi-neering group. One suggestion was for an inflatable vest that could be worn under an astronaut's pressure suit. By connecting it into the air-cleaner line, when air was in the lungs, it was not in the vest—and vice-versa. With the chest area of the suit always remaining the same vol-ume, it would further minimize the test subject's rise and fall in the water as they would breath.

Another idea was my answer to a problem that New York City posed about problems with aging dock facilities. Design a one-man sub for inspecting piers, ships, or other water structures. Such a vehicle could also be of use for other tasks—possibly furtive missions by the Navy. It would have a dry, air-filled cockpit or compartment for the operator. That person could don a mask and mouthpiece, let water into the cockpit, and exit the vehicle for close-up inspections. It would be a simple design to fill and empty the chamber, as submarines do with their ballast tanks.

Though I continued to look over my shoulder, and keep in touch, it wasn't possible to rejoin them. I was in the new assignment six months when GE announced that they would build an indoor pool in King of Prussia for future astronaut equipment test and training. How ironic! When the team was no longer together—now they'd build the facility we'd begged for. Fred and I were among those asked to contribute ideas on what should be included. And I'm forced to concede the final design was first class. Thirty by sixty feet in area, and thirty feet deep—an overhead crane for lifting equipment and personnel in or out. A dressing room with lockers and showers, plus a workshop adjoining the pool area. All of the things we could have used over the past three years.

Art Rachild was selected to oversee construction and become the eventual Supervisor for the facility—a wise choice. The surprise was that it wouldn't be built in one of the permanent buildings on the company owned hundred plus acres, or a new building. It was to be in one of the several metal buildings leased in the King of Prussia Industrial Park, on the opposite side of the Pennsylvania Turnpike.

Following the completion of Tektite, Fred and the others were assigned to a group that he felt had nothing to do with their work. I knew he was disgruntled, but a few weeks later I was shocked to learn that Fred, Carl, Jack Burt, Charlie Soult, and two technicians planned to leave GE. Fred talked with me before they resigned, but he realized that I had more years of seniority to sacrifice than they did, and a spouse who had not been either understanding or helpful in our ventures to date.

He and Carl had made contact with John DuPont, a wealthy man who had intended to play polo in upcoming Olympic games. But one of his required numbers of ponies went lame, and that forced him to look at other areas of interest. He provided three million dollars start up capital to the new company, with a proviso that they develop offshoot products from the original rebreather. They had to become self-supporting in two years. Du Pont had chosen his investment wisely. The rebreather principle was redesigned into a variety of hospital devices: an infant resuscitator, and an isolation booth for contaminated patients. The latter combined an air stream blowing from top to bottom inside of a Herculon enclosure. They teamed up for those products with the vendor I'd used for the huge shower curtain. Later, they reentered the business of diving equipment.

The MOL simulator cabin design was progressing well; a cylindrical structure with a plain exterior, it resembled a barrel set upright. Pierced in countless places by cables, the shell was surrounded by the support devices. Its interior was elaborate, complete with panels, switches, lights, and view ports fed by slide projectors. The 'view' in those windows would display whatever surface the vehicle would 'fly' over. Training for the Air Force astronauts was intended to give the impression of flight in every way except the presence of gravity.

Two MOL simulators were being built. One was to be set up at Vandenburg Air Force base in California. Halfway between San Francisco and Los Angeles that base covers a huge acreage; it's possible to drive all day and never leave the grounds. The other simulator would remain in King of Prussia as a backup. Having two units would also allow update of one cabin at a time, so training could be continuous. There would be no interruption of training during the periodic updates.

I was offered the opportunity to install the hardware at Vandenburg; it would take nine months during which I'd receive per diem pay. Then I could stay on site as an instructor for three years, with a renewable option. I'd visited the site, but hadn't agreed to accept the offer. Radios aren't permitted openly on the work floor, but many people keep one in their drawer to listen during work breaks, or at noon. In the simulator lab one morning in June of 1969, a technician called me over. He had a puzzled look on his face.

"I took a late break to listen to the news. They just said the MOL program is cancelled. Have you heard anything about it?"

"No one has even suggested it. Pressure was applied at the staff meeting this morning to meet the schedule revisions. Let me nose around."

"Okay," he responded, "but tell me if you hear anything."

I went back to the office and found Ted. He didn't know a thing about the newscast—or what their source was. Not having heard any more, I went to my car at noon. Sure enough the newscaster said, "The White House announced this morning that the Manned Orbiting Laboratory program has been cancelled." No other details were given.

After lunch the office was abuzz, but no one still knew anything firm. At two o'clock a memo was circulated. It told everyone to get any personal belongings or tools out of the lab by four o'clock—no exceptions. I had a file cabinet full of notes and drawings to move, and some hand tools. Then I helped the technicians roll their tool chests into the office; they didn't know where else to put them. By quitting time the office was jammed with people and equipment. It was the first time I'd ever seen some of them face-to-face. Groups talked together, but no one had heard more than rumor.

On the drive home that night, I thought about what to tell my family. In twenty-one years, this was as near as I'd ever been to a layoff, and it wasn't a comfortable feeling. I pulled into the garage. "Let's at least get dinner over with," I commented to myself, "before the family goes into a panic."

14

"Women are meant to be loved, not understood."

—Oscar Wilde, 1854–1900

My next-door neighbor was a manager for a different group at King of Prussia, and his wife called him the minute she'd heard the newscast about MOL being cancelled. Chris assured her that his work wasn't affected. Feeling secure, she proceeded to call every GE wife she knew and told them; mine was among those who got word in that fashion. By the time I arrived home she was near hysteria—fat chance of sitting down to talk things over.

What I did know, but didn't mention at home, was that five thousand MOL coworkers at Douglas Aircraft, in California, were sent out the door that day. They were handed a letter that effectively said, "MOL is cancelled, two weeks severance pay will be sent to your home address, don't call us—we'll call you." There were twenty-five hundred employees affected in King of Prussia, and I'm grateful that General Electric did better by their employees. With twenty-one years seniority it was my opinion that something else would open up. No worry. At least that was the approach I took at home: stiff upper lip.

The next day GE announced that those who had been hired in the last thirty days were to be terminated, but that impacted less than three hundred. To end expensive lease payments, we were told to pack the IBM 360-44 and Beckman hybrid computers, which had been used for the cabin simulator. Other equipment was dismantled for storage or reuse. A week later, those with less than ten years seniority were let go; longer service employees continued to dispose of hardware, files, and drawings. Unlike other projects I'd seen cancelled, nothing in the files was saved; everything on paper went into the red security boxes to be burned.

Two weeks later the company delivered layoff notices to those who had ten or more years of service, and yet another benefit was extended.

The office would be kept available; phones and secretaries were to be left there indefinitely. Resumes would be printed on request. Best of all, we'd receive a regular paycheck every week, though it was issued against the severance amount due each of us. Nonetheless, it was a generous move. Personnel who had been on loan to short-handed projects were returned to MOL as the other groups began to retain work for themselves. The impact of MOL's cancellation was spreading like pond ripples. Small contracts were also completed, and that added to the surplus of employees.

When things get bad, they often get worse before they reverse direction. My car fell apart. Before anyone suspected that I'd soon be unemployed I financed a replacement. The first thought, that another position could be found in King of Prussia or Philadelphia, proved to be wrong. Despite all of the resumes I submitted there was no offer—either locally, or with other divisions. I believe my resume received less interest because it showed a non-degreed engineer. My story for the family changed to, "There are fewer places to absorb everyone. We have to consider moving."

Ted Johnson had received an offer to manage a microelectronics unit in Syracuse, New York; before he left, it was his desire to see all of his people placed. He called our group together and said, "Think about what else you'd like to work on, and whether you would take a transfer. If so, where? I'll talk with each of you tomorrow. Maybe I can help find something that you're overlooking."

The computer-programming group of GE Missile Department hired Jack Billett; he was the only one of our unit who was able to remain in the area. Jack and his wife had their first child three weeks before. To work in Philadelphia and not King of Prussia made a longer drive from home, but they were pleased not to be forced into a move. George Fleming was the swinging bachelor of the group. He drove a Lotus, was an aficionado of modern art, and a computer whiz. In office discussions, his reference to the way he'd set up programming problems made me believe George thought of the machine as another person; he held conversations with them in his mind. It was his decision to travel in Europe for the third time. While there, he'd see if something were available. The last I heard of George, he was working as a programming consultant in Holland.

I talked with Ted about staying in computers—the mechanical end of the business. The manufacturing facilities for GE Information Systems were in Phoenix, Arizona, and Oklahoma City, Oklahoma. It was a plus that both cities are near Universities; wherever the next job was located I wanted to complete a degree and fill the void in my resume. Ted had worked in OKC before, but never in Phoenix. However, he seemed to know managers everywhere; phone calls opened doors for me at both places. I flew to Phoenix and interviewed with two managers who were working on a magnetic label reader system; TRADAR was a forerunner to modern price scanning systems, being developed exclusively for the J.C. Penney Company. It didn't use a bar code, but read labels that carried data on a one-inch strip of magnetic film.

Both of the groups offered a shared position to me, but not in design. It was to perform quality assurance—monitor parts made by outside contractors. Working for two bosses would be difficult, if not impossible, and they didn't need an engineer for the job. If I took that position I'd be on the road, or in the air, more often than not. Travel and time away from home had already created marital problems for me, but I wasn't in a position to refuse any employment and said, "I'll need to discuss the offer with my family, and let you know."

Arizona itself presented another problem—extreme heat. It was the end of June, and the airport temperature was 103 degrees by my flight departure time of eleven a.m. I had to walk out onto the tarmac to enter the plane by a stairway. As I left the shadow of the terminal building, intense heat struck me like a sledgehammer blow. Two hours later, when I arrived in Oklahoma City, it was a relatively cool 94 degrees. Talking with my spouse that evening I told her, "I'd prefer to find a position in design rather than quality control, and if OKC has anything at all to offer I'd rather live here than in Phoenix."

Oklahoma City wasn't where computers were built, just the peripheral devices—input and output hardware. Five lines were in production at the time: printers, disk drives, tape drives, banking terminals, and punch card readers. There were also sub-assembly lines for paper tape generators, read-write heads for the disk drives, printed circuit boards, and other small parts.

The next morning I met Lou Gitzendanner, manager of disk engineering; his superior, Ken Morrissey, was on vacation. In fact, few people were on hand. It was the annual two-week plant closing. "If I'd realized there was a shutdown," I told Lou, "we could have arranged my interview later. I didn't mean to cut into your vacation plans."

"I'm scheduled to work through these weeks anyway," he responded. "Don't worry about problems—you're here because I already have one. My last mechanical left six weeks ago. I need someone soon. Ted gave me some background on you, and thought you'd fit. I've read your resume, let's talk about experience."

I mentioned the Machine Design and Naval Ordnance work I'd done in Pittsfield; then spoke of more recent work on frame and structural designs. My sheetmetal and packaging experience seemed to be of most interest to him, and he mentioned a need for a new power supply design—right away.

"That's enough." Lou interrupted. "I'm offering a lateral move; the same level and the money you're earning today. It's really an increase of eight percent because of the lower cost of living here. If it's agreeable, I'd like you to defer any vacation time until we see daylight. When can you start?"

"I could stay now, or start next week—your choice."

"Glad to hear you say that. Stay until Friday; we'll get you through the paperwork in Personnel, and familiarize you with the products and buildings. Come back as soon as you can next week. Married? Any children?"

"Yes to both questions. A daughter eighteen, and a boy of six."

"Twelve years in between. Kind of like having two 'only' children isn't it?"

I grinned. "Sure is. I'll have to put in twenty-four straight years of PTA."

Lou picked up the phone, and dialed. "I'm busy this afternoon, but I have a friend in the real estate business who could show you houses in any price range you want. You may as well get familiar with what the area has to offer. Wives want to know things like that." When he hung up the phone he said, "We'll meet Harold for lunch at eleven thirty. Let's walk through the factory. I'll show you the products we build, then we'll look for Dave Chesher in Personnel."

He explained, "GE came to Oklahoma City in 1963. Senator Robert S. Kerr was on the Defense Appropriations committee and tried to bring interested employers here. MCD, Military Communications Department, was active here until the Senator died. When he passed away, so did his influence on contracts. When Information Systems Department in Phoenix sent the computer peripheral products here; they needed many of the same type workers we already had in OKC." Lou and I toured the engineering and assembly buildings.

At lunch he introduced me to Harold Jones, a real estate broker who owned his own agency. After he dropped Lou at the office, Harold asked questions about family size or special needs, and drove past a dozen homes in our price range. We walked through four empty homes. He knew that I wouldn't make a commitment without my spouse, but wanted me to see ceramic wall and floor tiles, woodwork, trim, and all of the things that were standard for homes in this area, not expensive extras. I called Ted Johnson, to advise him of my decision to accept Lou's offer in Oklahoma, and of my intention to stay until Friday. When I called home that night, it was a relief to tell the family where we would be living.

The next day I was surprised by a telephone call from Wayne Adams, one of my technicians in King of Prussia. He'd heard that I was going to accept a job in OKC, and called to relate a personal horror story. He said, "I was at Tinker Air Force base for two summers. No one could sleep inside. Our families took mattresses out on the runway, to get any breath of moving air. I thought that you should be warned."

I said, "Thanks for the concern, Wayne, but I've already committed. Besides, this is the end of June and the temperature isn't bad. Humidity seems to be lower than in Pennsylvania, and all of the houses I looked at are air-conditioned. I may have to get the cars updated with air conditioning, but that should be the worst of it."

The only other words he had for me were "Well—good luck. I'll see you when you get back here. If you haven't melted."

Friday afternoon I flew home, and returned with a larger selection of clothing on Sunday night. I wouldn't be moving the family and furniture until the house in Kimberton was sold. It proved fortunate that I'd started work that week. General Electric had begun to realize that larger layoffs were coming at Missile and Space Division; scheduled interviews were cancelled or delayed for weeks, but my records had already been transferred. Friends of mine with up to nineteen years seniority were let go because there weren't jobs to be found within the GE family. Most of the longer service people affected were non-degreed like myself. That reaffirmed in my mind that, come fall, I'd start back to class.

I didn't want to move to an apartment in Oklahoma City. We needed to buy a house, but my wife still didn't want to fly. She told me to pick a house, and she would "make do." I stood firm, and adamantly refused to buy one without her. When we'd moved from Willingboro, New Jersey, to Kimberton, I'd forced the move to shorten my driving time. She had reminded me too often that the split-level house in Kimberton, our home for the last four years, wasn't her choice.

She finally had to fly out or see the furniture go into storage. My parents drove from New York City to Kimberton to care for the two children. It was mid-summer; they weren't in school. Mom and dad also cared for our loveable watchdog; Duke always barked to let us know when there were strange noises, but would have licked a burglar to death if one came inside. A mixture of Basset Hound, Border Collie, and more, I've referred to the breed as Heinz 57, but it was the best dog we ever owned. One problem though—once six-year-old Rick called the dog Duke, she would never respond to Duchess.

Sheryl had just graduated from Phoenixville High. At the beginning of August, I had her fly out to take SAT exams at University of Oklahoma, in Norman, an hour south of OKC. She was accepted, and wanted to attend their Nursing program. I made arrangements for her to have another room at my motel until the O.U. dorm opened; at least that was the plan. My house in Kimberton sold, and the closing was scheduled. A week before that date there was a burglary attempt. Duke growled and barked when the rear screen door was jimmied. That woke my wife. When lights started to come on, the intruder was frightened away.

I'd taken Sherrie with me on weekend events: picnics, and boating at Lake Thunderbird. She had met Lou and Helen Gitzendanner; we'd been on their boat to water-ski. Lou ordered me to leave on the next flight. He said, "Have Sherrie pack. Helen will pick her up from the motel and have her stay with us next week." While the rest of our family was driving to Oklahoma, Helen played mother hen and settled Sherrie into her dormitory at Cleveland House, on the O.U. campus in Norman.

IBM built the first commercial disk drive in 1957, for their RAMAC 350. RAMAC is an acronym for Random Access Method of Accounting and Control. With a 5megabyte capacity, 50 disks of twenty four-inch diameter, and 100 read/write heads—it was huge. Companies could rent it for $150.00 per megabyte, per month. GE got into the disk drive business by buying OEM, or Original Equipment Manufacturer, equipment from Burroughs, Control Data, and IBM. There was an old model disk drive in the OKC computer room, when I first arrived in 1969. It was an early storage unit that had been produced in Phoenix sometime earlier. Six aluminum discs, three feet in diameter and one-half inch thick, were mounted horizontally on a vertical spindle. Sheets of magnetic material, like the tape used in a VCR, but larger, were glued in place. Although though no longer manufactured, it was then operational.

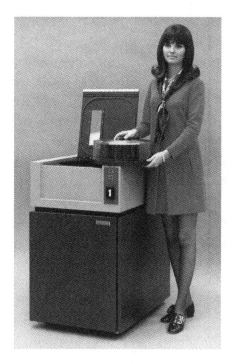

Disk Storage Unit DSU170B
Marketing Photo

The Disk Storage Unit that GE Information Systems Department had in production was a 250-pound model known as the DSU160C. A clone of IBM's model 1316, it had a 7.5megabyte capacity and an oil filled actuator that moved the read/write heads back and forth. The main deck had grooves around the actuator, to collect fluid leakage and channel it into a ten-cent tin can suspended below a drain hole; the quantity of oil that the hydraulic actuator leaked made the can an essential part of that design. Unlike a drum, which uses a permanent surface to store memory, the 160-Charlie used removable media. Early disk packs contained six fourteen-inch diameter disks inside of a removable plastic dust cover; for cleanliness, read/write heads never accessed the top and bottommost disk surfaces because of the possibility of dust particles.

A media pack was placed on a spindle in the front chamber, its cover removed and the lid was closed. Service access to the actuator was by raising the top hood section, just like raising and propping up the hood of a car. Power supplies were accessed through the short front door, and the printed circuit card cage was reached through a similar rear door. A similar model, the DSU170B was in the design phase when I arrived, but it never reached production. It carried twice the number of disks in a pack, and doubled the areal density written on the magnetic media from 100 bits per inch to 200. That would quadruple the pack capacity to 30 megabytes. Those two models were the state-of-the-art when I began to design computer peripherals.

New ideas were openly sought, listened to and appreciated. Personal satisfaction builds when others use your ideas; it creates self-esteem. Again I'd entered a field so new that the work could stimulate and excite everyone. It was an age when an inventor walking down Wall Street could yell "computers" and have money thrown at them. One of the other things that I learned was not to call the punch cards IBM cards. IBM may have made them better known, but they weren't the originators. They are Hollerith cards, named after the man who created them for the 1890 census of the United States.

Nine weeks after I'd arrived in Oklahoma City, the Tradar program that I would have worked on in Phoenix was cancelled. There were technical difficulties that may or may not have been insurmountable, but J.C. Penny opted to stop funding it. Several weeks earlier, the system locked up all of the electronics involved. That doesn't sound too bad when you say it, but none of the J.C Penny's stores in the Los Angeles area could open their electronic cash registers that Saturday. They sent their retail clerks home at two o'clock in the afternoon. Loss of weekend sales revenue may have been a deciding factor; nonetheless, had I taken that job in Phoenix, I'd have been involved with another layoff that soon.

Since no paperwork had been preserved at the MOL office, I returned to King of Prussia two more times. There were phone calls from individuals that were trying to install equipment whose drawings and memos had been destroyed. For security reasons the information couldn't be discussed over open phone lines.

In getting Sherrie ready for college, I discovered that Engineering classes at OU and OSU are scheduled in the daytime—few were offered at night when I'd be free. Though burned as I had been before, it was still easy to overlook small, important details. But I learned that Central State College, in Edmond, Oklahoma, had more commuter students than were living on campus. CSC offered a variety of evening classes—not in engineering—but they offered a Bachelor of Science degree in Computer Science. I had acceptance as an engineer through my PE license, a BS in a related science would be complementary to my engineering credentials. After talking with an advisor, I took their catalog home and made a list of all the classes that were needed.

In the spring of 1970 I worked on the next model of disk drives, I began redesign of the disk drives which used a hydraulic actuator; that unit was to be replaced by an electronic or voice-coil actuator, plus other internal modernization. The cabinet design remained the same.

Electronic Actuator Flying Read/Write Heads

My engineering supervisor was Simon Peter Fleming. Pete and I hit it off well. He found out that I'd lived in the Sheepshead Bay district of Brooklyn. At one time there was a racetrack located there. His father had been a horse trainer, and had traveled with his teenage sons during race seasons. The track was defunct by the time we met, but Pete remembered and liked the area.

Our design modification was completed just as a high-level decision was made to sell the department. We were all directed to be in one of the three cafeterias, at three p.m., there were too many employees to fit into one building for a meeting. No one had ever called simultaneous meetings before. People were standing in the rear, and the crowd wouldn't quiet down. Frank Lenher, the General Manager, tried to read from a paper he was holding, but the buzz was so loud that few heard him. He stood, silently, waiting for the audience to grow quiet. Gradually they did, and someone called out, "What did he say?"

Pete Fleming's voice boomed over the crowd, saying, "Massa done sold the plantation!"

Humorous as his words are today, they weren't then. The now silent crowd listened as the prepared statement was re-read. Details were reported later in Fortune magazine. GE needed to enlarge its market share in order to be more efficient. They'd studied other companies and picked Honeywell as the best candidate. There was little overlap in the product lines, and GE planned to pump $250 million into new products. But the U.S. Justice Department said no. Between the two companies they sold eight percent of the market. GE asked how that could be a problem, and were informed that they had too much financial clout to be allowed the purchase. Their buy position was reversed to sell; GE kept the Information Systems name for the service organization they would continue, but manufacturing of the computer and peripheral products went to Honeywell. By the time of that sale in October 1970, the department was producing 46 drives per week.

Honeywell Information Systems, Inc. was formed as a division under Honeywell, Inc. They took active management of the Oklahoma City and Phoenix plants on October 1, 1969. It proved an easy marriage; only thirty-seven people chose not to join HISI and were quietly moved out of the plant. Structure of the organization hardly changed. A few high-level staff from Honeywell showed up, but for the most part it was business as usual. Most of us did the same jobs, for the same managers.

Our new employer did want to move small independent groups to Oklahoma City, but it was difficult to convince people from Billerica, a suburb of Boston, Massachusetts, to consider a move to the rural Midwest. Twice the company hired a charter flight to bring Massachusetts employees and their family members west for a weekend. Housed in several motels, they were given tours hosted by local employees, realtors, and clergy. The visitors were escorted through Universities, the Civic Center, and on a tour of employee homes to see the quality of life that was available. It was necessary to convince them that the area housed culture, and that we weren't fighting Indians over the back fence like the movies portray. What we actually proved was that the

Indians had civilized the Territory before they sold pieces of it to the white-eyes. Over the next six months, small units transferred their programs here, and employment reached as high as 3,600 people.

Real estate prices were high in Massachusetts compared to Oklahoma, and most transferees needed to continue deferral of federal income tax on their personal homes. They sold older homes at top prices, and bought more house here for the money. Half of those who finally moved to Oklahoma bought RV's and boats with capital they held back from their new down payment. Six months later, some wag started a rumor that Non-Impact Printer group was to return to Billerica. The result was a near rebellion. Few wanted to leave.

My first assignment under Honeywell was to fit the GE redesigned DSU160C and DSU170B voice-coil models into a cabinet that was as close as possible to the Honeywell exterior. Parts relocation and bracket modification made for a complete cosmetic change. New sheet metal side skins disguised the old chassis, but cabling to the new family of products also had to be common. I created a sheet metal adapter to allow the new style of entry cable connections. That cable adapter method was used on all existing hardware, and led to my designation by Ken Morrissey as hardware coordinator for the five local products; occasionally I traveled to Billerica, Massachusetts, for review of our compatibility with home office designs.

DSU160C/DSU170B
GE Products in a Honeywell Style Cabinet

My immediate manager, Larry Root, came from Billerica and was of great assistance to my completion of CSC classes. Four of them could only be taken during the day. Twice I had to attend class from eleven to two o'clock for a semester, and Larry adjusted my work schedule. A minor in mathematics presented a small dilemma. I thought I'd transferred enough classes to cover it, but Drexel evening college had conducted three semesters during their school year—a trimester schedule—whereas Central State had two per year. CSC had to prorate Drexel credit hours, and I was one-third of a credit hour short from completion of my math minor. It was necessary to take a math course from the third or fourth year to finish the requirement. I found a suitable course offered in the next semester, and immediately enrolled. It had been twelve years since I'd taken any mathematics at all, and given little thought to needing a refresher. I scheduled a semester of Differential Equations, back-to-back with a class in Statistical Methods II. There was an unbelievable amount of paperwork for those classes, and it was the hardest time in my student life. I'm not sure how, but I made it through.

By the time I received my Bachelor of Science degree in May of 1973, Central State College had been renamed Central State University. Mother and dad were still in Brooklyn. They flew for the first time when they came to my graduation. It had been twenty-five years coming and they considered it an important enough event to make the trip; unknown to me, they had another motive. Dad had been ill for ten days during the previous winter, and mother couldn't get him to the doctor's office. She held a valid driver's license, but hadn't driven a car for years; she only used her license as identification to cash checks.

Their doctor, fortunately a good friend, had made several house visits to see him. When dad recovered, dad asked mother, "Why are we living here, fifteen-hundred miles from either one of the kids?" While they were visiting in Oklahoma City, there were thunderstorms and tornadoes. Next they visited Warren and his family in Miami, Florida. I thought they move to the sunshine state, but at age 67 and 77 respectively mom and dad chose to live in Oklahoma. They bought a small ranch style house ten blocks from my home in the OKC suburb of Warr Acres.

Later dad told me," In Miami there wasn't a house within ten miles of Warren's place that we could afford. I wasn't worried about weather, but Florida would make me take eye exams annually. You see a lot of people on three-wheeled bicycles there, and my backside isn't shaped to fit the seat on one of those damn bikes." Dad never put ten miles a day on the car—to take mother for lunch, buy groceries, or a doctor's visit—but independence was important. They settled in with the knowledge that my division had been sold once, and relocation was always a possibility.

Honeywell had previously designed their own hardware, had the parts made by outside contractors, and assembled the units in-house. There wasn't much change felt at Phoenix or Oklahoma City because our new owner didn't have the experience with in-house manufacturing that we did. If it isn't broke don't fix it; in our case, don't change it. That philosophy held for five years, until Honeywell decided they'd like to return to the old way of buying, not making, their hardware.

HISI merged in 1975 with Control Data Corporation; it was a marriage of convenience for Honeywell. With ownership of thirty percent Honeywell, and seventy percent CDC, all manufacturing operations were CDC managed. Magnetic Peripherals, Inc. was formed, MPI for short. We became an Original Equipment Manufacturer. Finished products were provided to Honeywell Computer Department and anyone else that wanted to lease or purchase our products.

Ten employees were scheduled for relocation to the Honeywell facility in Billerica for product service—not in design, but as continuation engineers, or troubleshooters for existing products. My seniority of twenty-seven years put me in consideration for that list. Ken Morrissey called me into his office. I thought he wanted to ask my choice, but he threw a surprise. "This is strictly between us, but with a little luck things are going to get better here. I want to ask you to do something for me."

"What do you need, Ken?"

"You're on the list to move to Billerica, but I want you to remove your name from consideration. There's no guarantee that anything else will happen, it's a gamble. But I know you aren't pleased about the prospect of moving back to Massachusetts."

"I've never had a reason not to trust you, I replied. "Take me off the list." I never told my family of that choice, but there was never reason to regret it. A week later Ken and his manager, Dean Bowman, called seven of us to a conference room.

Dean said, "CDC has an offer from Northrop Aviation, for a building they own in Watts, California. Our Board of Directors has agreed to move the products to OKC, but they haven't negotiated a price for the building yet. We can't afford to lose weeks doing nothing. I want you to do everything to make that product transfer work, without making it known in the meantime." The seven of us were moved to a locked room setup, where we could see and handle drawings or hardware, our own and competitor products, without being observed by others. It was a Kelly Johnson Skunk Works type of operation.

CDC-Watts built two types of drives, and those product lines were now scheduled for a move to Oklahoma City. One was rigid media drives that included a product with a removable heads and media package, a technology known in the trade as Winchester drives. The other drive was a new type using a media called a flexible disk. The product had started as a read-only device, installed in the multi million dollar computers as a diagnostic tool for service calls. Next had come the read/write version, but it still only wrote on one side of the diskette.

In the fall of 1975, Watts' floppy disk production was in its infancy; building drives for eight-inch flexible disks and yielding twenty-five units a day. It was my responsibility to work out the remaining engineering problems so the floppy could be produced in Oklahoma City. I tried to convince management to change from sand castings to die castings, which are more precise in dimensions. It would have solved many of the production problems, but required a large expenditure for the molds.

I was told, "Forget it. How many of these things do you think we'll ever sell?" By spring 1976 an assembly line was established in Oklahoma City that turned out 18,000 floppy drives in nine months, and production increased dramatically every year after that. Next it became possible to write both sides of a diskette, and that improvement was followed by the disk's density being doubled.

Engineering Test Lab
8 inch Floppy Disk Drives

The next generation, 5-1/4 inch drives, were built using the die cast frames I'd touted for accuracy. There is an interesting reason why the next drive size chosen was an odd number like a 5-1/4 inch disk. It was to be the replacement for tape cassettes, which numerous computer sellers had designed into their hardware package, panels, desks, etc. It had to fit the same space as the recorder with no machining. A few years later, foreign competition and cost-cutting forced the company to stop manufacture of the floppy disk line. A floppy can be bought today for as little as twenty dollars

At the time classes had been selected for my BS in Computer Science, I'd used business courses as electives; prerequisites for a Masters in Business Administration were met that way. An additional thirty-four more credit hours allowed me to complete that graduate program as well. In May of 1977, I received a Masters in Business Administration from the University of Central Oklahoma.

When we'd first relocated to Oklahoma, and all that we as a family held dear was under threat by circumstances beyond our control, the family strife had ceased. For five or six years, everything at home seemed to improve, but it was the calm before the storm. Since my arrival here, I'd traveled very few times. More recently, as Honeywell merged their interests with Control Data Corporation, I'd traveled four times to Watts California for a total of nine weeks away. On one of those trips I'd brought my wife and son to California for a week, to visit Disneyland, Knott's Berry Farm and other tourist sights. But travel stopped being the biggest complaint; now it was school attendance, and time for homework, that became the focus of renewed criticism by my spouse. Though friends and neighbors weren't exposed to the cracks that were building, it seemed I could do nothing right anymore. We had shared good times and bad, but now my home front was truly under siege.

I concede the absences from home, and late nights for school, but it was the continual upgrade of education that which improved my performance on increasingly technical work. And that's what enhanced our financial position and lifestyle. Nevertheless, married life seemed to be over. Sheryl had married and was long gone from the household. Rick was twelve years younger than his sister was, and I procrastinated in any action to see if things would be resolved when school was finished. That was a vain hope. In April of 1978 it became necessary for me to leave. Sherrie saw the situation in an adult manner. Rick at sixteen took it very hard.

15

"Happiness comes through doors that you didn't even know you left open."

—On the Internet, author unknown

Dating was something I hardly expected to do again in my lifetime, and I did not do any for two years; however, I'd been married too long to become a hermit or want a solitary life. I enjoyed dancing and, in Oklahoma, boot scooting is done everywhere. The dance floors are full of couples doing line dances, or the cowboy two step; square dancing can be found everywhere. Ladies that I dated were always someone with whom I would consider marriage. When that certain someone crossed my path, there wasn't any hesitation on my part.

Wanda and I have so many things in common it would make a long list. She was a single parent with two children: Carol in college, Stephen a senior in high school. Until then, she was employed as nurse and office manager for a local surgeon; now, with financial and moral support, she seized the opportunity to complete her Bachelor of Science degree in Nursing.

Honeymoon in Dallas
July 6, 1981

In 1985 we opened a home health agency called HomeCare Nursing Services; the office was in the medical building across from Edmond Hospital. Wanda ran the day-to-day operations, but as her Assistant Administrator it was a great opportunity for me to make use of my MBA. I continued working for Control Data, but nights and weekends I paid taxes, wrote payroll checks, and filed the annual eighty-page form that was submitted to Health Care Financing Administration for cost verification. The intermediary used for Medicare payment was Blue Cross/Blue Shield of Iowa. We traveled to Des Moines for training on how to fill out and file their forms, et al. The rules to satisfy HCFA are complicated, and expensive for a mom-and-pop type agency; they show favor to chain organizations, or hospital-based groups. At the end of our training I asked, "What else do you think we might need?"

The response given by their head of accounting was "Good luck, and deep pockets."

He was right. Being allowed to bill only once a month, plus their time to review cases and respond with more questions before payment was received, there was often a need to obtain money. Some rules were ridiculous and only obvious if they were suddenly applied to the agency. For example, HCFA allowed interest on a bank loan to be listed as a business expense, but not if it was the owners who loaned money to their own company. We eventually faced the problem that made GE sell the computer division—get bigger or get out. We closed the office at the end of 1989.

Wanda's primary interest in nursing is the elderly—gerontology. Whenever a University class was offered in a related subject, she enrolled. Before long, she had enough credits for a Masters degree in Education. After closing the agency, she was Director of Nurses for several nursing homes and a Hospice group. Eventually Wanda became a nurse educator at Platt College. She is still active with OGNA, Oklahoma Gerontological Nurses Association.

Control Data Corporation manufactured a vast line of disk drives. As long as there was sales interest in sufficient quantity to be profitable, a line was continued. Otherwise the product was retired. Mainland China was looking for computer products they could build, to modernize their business systems. Some equipment that the company made was too advanced in technology for the U.S. government to allow export, but there were older designs that CDC no longer built. Mainland China expressed interest in buying the rights, tooling, and details of the manufacturing process for one of the older disk drives.

Oklahoma City became host to eight of their engineers so they could study the product itself, and the manufacturing procedures, before they finalized an agreement to buy the line. They were housed in the motel where I'd stayed when I first arrived in Oklahoma City, the Tropicana Inn, on NW 39th expressway near the corner of May Avenue. After changing hands several times, it is now a Day's Inn franchise. Two weeks after their arrival, a drunken motel guest emptied a revolver at something on the wall that he considered offensive. Next morning one of the Chinese engineers showed us three bullet holes in his suit. He was understandably excited. His tale was punctuated by "Bang. Bang," as he shaped his finger and thumb to portray a pistol.

Their interpreter explained that, when they left home, all of them were supplied western dress by their government. To keep it neat and clean, it wasn't worn inside the motel room. His coat and pants had been hanging on the wall when the bullets came through. Ken Morrissey had offered to replace the garment, but the interpreter explained that the engineer wanted to convince his government to let him keep the suit as a souvenir. He wanted to show friends at home what had happened to him in the Wild West.

I moved from Product Design Engineering to Test Systems Engineering, a group of seventy-four people involved in the design and manufacture of automated test and assembly equipment that supported the product lines. TSE produced laboratory-grade, automated machinery. When I first joined the unit, I shared an office with Jack Breimeir, a technician with whom I'd work; I was designing a read/write head tester. The head carriage was moved into position by a motor driving a screw-shaft that looked like a 'Yankee screwdriver'; it delivered a rapid advance. But during the test phase, the head was moved by a different screw and motor to produce 125 microinch steps. Jack watched as I used the top of my desk to create sketches, and order parts. When the piece parts arrived, he helped me to assemble them, install motors, add computers, and tie it into a neat package.

He said, "I've never understood how people can start with a blank sheet of paper and create machines. It has to be a natural gift."

I told him, "It gets so you can see them in your head before you put it on paper. Perhaps that's my gift—I picture what is needed. You have a different gift—to assemble them, create the right fit, calibrate twenty machines that produce identical results, and give me feedback to update the drawings."

Honeywell Information Systems, Inc. merged its interests with Control Data Corporation, in 1975; over their eighteen years of operating the Oklahoma City plant, CDC bought back the shares held by Honeywell. Disk drives and other peripheral products continued to shrink in size, while multiplying their capacity.

Wren 5-1/4 inch Hard Drive

The late eighties saw networking of desktop computers lead the industry in another direction. IBM, Control Data, and other large-frame manufacturers were taken by surprise as sales of multi-million dollar large frame computers came to a virtual standstill. Worse yet, the purchasers found ways to stall their contracts. If a buyer killed their order outright they would have to pay cancellation charges, which could be considerable—instead they would postpone delivery of the unfinished unit. CDC's loss for 1987 was five hundred million dollars; in 1988 the red ink stopped at one hundred and fifty million. In early 1989, I flew to CDC headquarters in Minneapolis, Minnesota for a cost improvement award presentation, and banquet. Part of the trip included a factory tour. Incomplete frames stood like skeletons on the factory floor, it was a walk through dinosaur land. By year's end 1989, the company had lost another five hundred million.

For employees of Oklahoma City Operations

Volume 20, Number 5   May 26, 1989

Imprimis delivers world's first 5.25-inch 1.2GB drive

The world's first 1.2 gigabyte disk drive in the 5.25-inch form factor was hand-carried to Zenith earlier this month.

"It was like watching a kid playing with a new toy to see Zenith officials marveling over the large capacity and performance of the WREN VII," said Steven Genheimer, Advanced Design Manager, one of three Imprimis representatives who delivered the drive to its first customer.

Account Representative Randy Lee and Product Line Manager Marty Christensen also shared in the historic delivery.

"To say they were impressed would be putting it mildly," Genheimer said. "It

was real exciting, as well as nerve-racking, to be a part of the plug-in-play of the first 1.2GB in the industry. They loved it and so did we. The drive came through just as planned."

Genheimer, who also served as manager of the WREN VII development team, said Imprimis gained a lot of credibility within the industry by delivering the new drive on time.

"Yes, our reputation was on the line," agreed Dilep Patel, Director of FII Engineering. "We announced the development of the 1.2GB at Fall COMDEX with a delivery date near the end of the first quarter. We're right on schedule."

One thing that's ahead of schedule was the announcement of a 5.25-inch drive able to handle such a large capacity. "We usually announce a product after development, not before," Patel said. "But this time we announced the 1.2GB before we had drives ready to ship because we wanted to establish the industry standard in this form factor," he said. "Which we did."

"There's still a lot more work to do, but this really boosted our confidence that we can achieve even higher capacity in the same form factor," Patel said, adding that "we're already looking ahead."

Genheimer expanded on Patel's comments saying this is the first time Imprimis has been the leader in defining a brand new product, "We've been playing catch-up in the past in the development area while leading in other areas of quality and performance."

The WREN VII program team gives the "thumbs up" sign in honor of Imprimis' development of the world's first 1.2GB drive in the 5.25 form factor. Team members include Pete Flemming and Steve Genheimer, top, and Bill Stubblefield, Tom Lemming, Garry Liehl, Steve Welty, Buck Pollsell, Bill Boudreau, John Worden, Chris Lagaly, Marty Christensen, Mary Blankenship, Darrell Sedbrook and Andy Wakeland.

Oklahoma City Operations Vice President Tony Maggio said it was a real distinction to work for a company that produced the first 1.2GB in the world in the 5.25 format.

Before the new Imprimis drive hit the market, the largest capacity drive was 760MB.

"We feel this large capacity drive shows clear leadership in the marketplace for both Imprimis and Oklahoma City," Maggio said.

"We are getting excellent response on this new product, which is the culmination of continuous progress in performance and capacity," he added.

Marty Christensen, WREN VII Program Manager, said it took the effort of

several people to achieve this milestone. "Customer reactions worldwide are amazing. This program clearly has the potential for significant positive results for all of us," he stated.

Less than a year in development, the new drive was designed by a 12-person development team under Patel's direction. "There was a lot of excitement along with the hard work. The whole team from engineers to manufacturing to test came through in the clutch," Patel said.

In addition to Genheimer and Christensen, the program team consists of Steve Welty, Darrell Sedbrook, Chris Lagaly, Bill Boudreau, John Worden, Garry Liehl, Tom Lemming, Jack Hayes and Bill Stubblefield.

Tony Maggio, Oklahoma City Vice President of Operations, enjoys visiting with those who helped Imprimis produce the world's first 5.25-inch disk drive with a capacity of 1.2GB at a special luncheon held in their honor.

Announcement of the World's First
Gigabyte Capacity Disk Drive
May 26, 1989

When the disk drive operation in Oklahoma City and Minneapolis was renamed as Imprimus, which means "In The First Place,' handwriting was on the wall. Our department was being severed from CDC so it could be sold. March 1989, while known as Imprimus, the world's first gigabyte disk drive was announced by Oklahoma City; industry had needs and we were still filling them. On October 1, 1989, the facility was sold—to a small disc manufacturer named Seagate. Financial and industry pundits joked about the tail wagging the dog. As time often proves, it was good for both. Seagate became the world's largest manufacturer of disc drives running the gamut from cheaper gaming drives to the most sophisticated in the business.

For me, another problem arose. In order to preserve my retirement benefits that included medical insurance, I had to take early retirement from Control Data. October 1989 saw my break of employment with what I always referred to as, 'The Company.' At age 59, I chose not to look for another engineering position somewhere else, and decided to work for myself. The impending sale had been obvious, and I'd had time to prepare. Dave Haggard, supervisor of a regional office for Farmer's Insurance, taught four sales rookies three nights each week. When we felt ready to take the Oklahoma licensing exam for insurance sales, we could do so and work full time with him.

I received my license before my retirement date and established a part time presence at Farmer's regional office. The best investment you can make is in yourself. When another agent left, I purchased his hand-held computer and other office equipment. Retirement from CDC merely changed the office where I worked every day. To keep in contact with friends, I visited them at the plant or met them for lunch. At first they were concerned that I'd be haunting them about their insurance needs, but I'd reduced living expenses—we could manage on my pension plus the small income from Bishop Insurance Agency. It wasn't necessary to push insurance at friends to earn a living.

When they no longer felt pressured, those people started calling me. "How about giving me a comparative quote on my car or house insurance?" Sometimes they'd call for advice. "My son turned sixteen and got his license yesterday. What should I do about adding him on my policy?"

I'd tell them, "At sixteen it isn't a question of whether they'll have an accident, statistics show that it's only a matter of time until they have one. I think that a driver who has his own money in a car is a little more careful about what they do to it. But, if you're providing a car for them, buy an old Buick or Oldsmobile—something built like a tank that will provide steel all around them. Make them pay part or all of the difference in your premium, so they become aware of what their driving privilege costs."

Four months later there was a surprise call from Charles Bussert, the manager for whom I'd last worked at Control Data. He hadn't found a replacement for me, and wanted to know if I'd be interested in coming back as a sub-contractor to help out for a few months. I agreed to work seven a.m. to three thirty p.m. for Seagate, and sell insurance on nights and weekends. I was employed through a temporary-staffing agency known as Defense Technologies, Inc., and sat at the same desk I'd left.

In mid-summer Charles asked, "Would you accept a permanent position again, if I can arrange it? If so, I'll get an offer pushed through Personnel."

The company was always slow in making offers. I've known people who accepted other jobs, and worked a month elsewhere, before getting a formal job offer from Personnel. I told Charles, "I'd love to be back working directly instead of temporary status, but you have to get an offer to me soon. Farmer's has scheduled my final training class at Kansas City, in July. Whichever deal comes through first is what I'll have to take."

Following the Farmer's class, insurance work would become my full time commitment. I'd move from the regional office, hire a secretary, and set up my own office. The training class scheduled in July had to be postponed—there weren't enough new agents ready; the next one would be in August. Charlie managed to push the offer through before then, and I accepted. Another new and upcoming agent purchased all of my insurance paraphernalia. To tell the truth, I was ecstatic. It was a lot of work to have two jobs when you're supposed to be retired. Leaving the insurance business was my second retirement.

It was an era when Computer Assisted Design, or CAD, was coming into common use. Software companies offered systems that would produce simple line drawings, or others with so many bells and whistles that three months of training were required. Workstations and licensed software became major investments—thirty thousand dollars or more to equip each draftsman or engineer. The total runs up, and management has to budget scarce dollars. Charlie had to decide how much he would allocate to modernize my equipment. Not knowing when or whether I might retire again, he upgraded my computer and software, but didn't provide a workstation.

I was the last person at Oklahoma City Seagate to use a drawing board for more than a book holder. My hand-drawn or red lined prints saw their way into the hands of contractors who made parts for us. Those drawings were eventually given to a drafter, who would enter all of the piece parts into a database; many such parts, drawn in layers, produce an assembly. When my intermediary in drafting completed the database, their workstation could spit out information to the printers or plotters faster than I can blink, and give dimensions to whatever number of decimal places is requested—three places, six, or more if that would ever be needed. But until they got all of my drawings entered into the database, I was faster.

If a new product came out and needed a different cradle to hold it, or a clock track was to be written by having the clock head enter a tiny hole in the side of a casting—any change at all. I could produce redlined prints, new part drawings, dimensioned sketches, whatever it took to build a part. Quite often, first hardware pieces for a new design were ready to assemble before the computer database was complete.

When Seagate decided in 1995 to have commonality in the way that our Servo Track Writer's wrote tracks on a drive, no matter which plant a disk drive was formatted, it took me a year's work to reach that goal. After the purchase of Conner Peripherals in February of 1996, Charlie told me that the company now wanted to include Conner's drives in that commonality.

It was more than I wanted to tackle again. "That's too much work to begin all over," I told him. "They say the third time is the charm. I'm going to retire—this time for good."

Once again Charles took a long time to find a replacement. To help Charles until he had a replacement for me, I phased into retirement. Two or three days a week I would troubleshoot problems with current production. On the other days, I took off, at least until my vacation time ran out in July. It was August before my replacement became available. The chap who did accept made it a condition that he'd be given the workstation and training that I wasn't able to talk them into. But, my employment had been a good arrangement for both the company and myself over the final six years.

I told Charles, "If I'd realized that it was this hard to locate another mechanical engineer in Oklahoma, I'd have asked for a big raise a long time ago."

Now able to travel, Wanda and I began a sightseeing trip along the Gulf Coast, to an eventual destination of Pensacola, Florida. Sixty miles east of New Orleans we crossed over the Louisiana-Mississippi state line on U.S. highway 10. At a tourist rest center I found a brochure for John C. Stennis Space Center, which three decades back I had known as Mississippi Test. Since it had been part of my first assignment as an engineer, I wanted to see how it had changed since then.

The Hall of Achievements showcases the American saga of rockets and space. Photos, actual flight hardware, and dioramas tell step-by-step stories of Mercury, Gemini, Apollo and men on the Moon, Skylab, and Space Shuttle missions. There were demonstrations, exhibits, films, and lectures on base activities. Inside the lobby is a quotation appropriate for the effort of everyone involved in space activities:

> *There is always the thorn before the rose.*
> *You have got to make some sacrifices,*
> *but you will be taking part in greatness.*
>
> —Senator John C. Stennis

It is NASA's largest site, and today the base is shared. They continue to test rocket motors for their programs. NOAA does weather tracking and prediction. The U.S. Navy operates Naval Research Laboratory, Naval Oceanography Command, and National Data Buoy Center; it is the only non-military site that maintains a full admiral in residence. National Marine Fisheries, U.S. Geological Survey, and Environmental Protection Agency are also represented.

One stop was in front of an observation tower—an office building on a sixty-foot pedestal is a more apt description. The tower is situated in a forest clearing, surrounded by radial paths through the tree line. At the far end of each bare strip, a mile away from the tower building, are the test stands. The tourguide, a young lady in her twenties, explained, "This tower is the original position from where each firing was observed by using telescopes." She drove to another parking space, and positioned us one hundred yards from a test stand, and said, "This is where we watch from now."

In front of me, an S-1 rocket stood upright. It wasn't a complete missile, just the bottom motor-stage. At thirty-three feet in diameter and one hundred sixty-five feet high, the top loomed over the stately pines that surrounded it. I had the momentary feeling that it was scanning the forest; there was a sense of awe, that the S-1was watching me and not the reverse. When all three motors of Saturn-V are assembled with Apollo on top they rise a majestic three hundred sixty-four feet. Each motor is tested separately; S-I is the largest, nearly half the height of the whole system.

Though huge in its own right, the S-1 was dwarfed by the structure it rested upon. Larger than two football fields at the base, and a towering twenty-four stories high, its work platforms reached out to entwine and embrace the missile like the arms of a lover. Technicians swarmed around the vehicle, performing checks required before a test firing. Hoses were dumping vast quantities of frozen, gaseous fuel to the thirsty tank.

Vapor clouds, visible at the hose connections, caused a younger tourist to ask her companion, "Is that steam?" "No," the man said. "It's LOX—short for liquid oxygen."

I chuckled. Her comment made me conjure up a Jules Verne-like vision of a steam-driven rocket being launched to the moon.

Then the guide added what had to be an unscripted, personal comment. "I don't know why they ever used the tower—you can see so much better from here."

It wasn't said to make her sound foolish—the words slipped past my lips without thought. "Because we blew up a lot more of them in those days."

Some on the bus snickered or laughed outright. A few caught the serious side of my comment. They were of sufficient age to remember when the U.S. saw failure as well as success. Today it would be difficult to find someone for whom the sights of a missile launch, or the return of a Space Shuttle, are not familiar. Component miniaturization allows larger payloads. The International Space Station, Hubbell Telescope, and Martian photographs are spectacular. But the comment made by our tourguide reminded me that there are many people who have never learned, or don't remember, what the pioneer days of missiles and space were like. Days when no one knew for sure how to take the first step to accomplish something never tried before.

President John F. Kennedy told us what we needed to do to serve science and help mankind, as well as recouping American pride. For the Mercury, Gemini, Apollo moon landings alone, there are 400,000 stories.

16

"And now, the rest of the story...."

—Paul Harvey

The Garden of Eden in my memories of Brooklyn doesn't exist any longer. During World War-II builders were under pressure to provide more housing. New York City Transit Authority sold the land adjacent to their train tracks; those treasured slopes were converted into row-style town houses. My Eden is forever buried in concrete. In time, nature showed that it has a way of getting even with mankind for trespass. Two years after those homes were built, a major winter storm struck the city. Plows attempted to keep the snow buildup from hampering travel, but they had nowhere to pile that much of the white stuff. Transit Authority didn't own land beyond the chain link fence any longer—they couldn't push it onto the patios behind the condominiums.

Express trains using the center tracks were halted, and mounds were created there, but snowfall continued. To prevent paralysis of the transport system through that section, train cars were lined up for seven miles from Coney Island to Prospect Park, where tracks go underground. If a commuter could reach a station they could at least walk through the cars as far as Prospect Park and ride in the tunnels beyond there.

The Authority deserves credit for what it does accomplish. They keep services operating for one of the major transportation systems of the world. In the year 2000 their trains logged 940,000 miles—nearly twenty-four times around the world. But a trip doesn't cost a nickel anymore; it's a dollar and a half for a one-way ride.

Baseball has returned to my hometown after its long hiatus. Not a major league team and not at Ebbett's Field; that hallowed ground was converted for apartment houses long ago. But on June 25, 2001, the Brooklyn Cyclones, a short-season class-A affiliate of the New York

Mets opened in Key Span Park—a new oceanfront stadium in Coney Island. The team is named after the seventy-five year old, wooden roller coaster that operates nearby on Surf Ave.

My older brother, Warren, had left his senior year of high school to enlist in the submarine service. A sidelight of note is that the two years of German language studied at Brooklyn Tech later qualified him for special assignment. The Navy wanted to test features on captured U-boats without the need to change machinery nameplates. He became part of a prize-crew operating the U-3008 for his last six months of duty. During Caribbean maneuvers for President Harry S. Truman, their boat was sunk in ninety feet of water when a practice bomb dropped too close. However, they were able to surface within two hours with no outside help.

By serving before the war was officially declared as ended, the GI bill helped Warren earn his baccalaureate in Accounting. He was able to secure his degree years before I managed to do so. Later he returned to school for a Masters in Finance, and worked for several large New York City accounting firms until his family moved to Florida. He was chief finance officer for several school systems in that state, and settled into early retirement in Tallahassee.

In November of 1944, when Lieutenant John Bishop, was killed, the 254th Marine squadron was preparing to go on leave. It was to be their first Christmas off duty since they'd arrived in the Pacific. Pilots put in cockpit hours that would qualify them for flight pay while they were gone. John was flying rear seat, with Lieutenant Allen Barber at the controls of a two man aircraft when the plane didn't pull up from a dive. Mechanical problems were an obvious suspect, but the plane sank rapidly. In 1998, I was invited to be a guest at a squadron reunion in Pensacola, Florida. Among those present were two of the four other pilots who were in the air that fateful day. They told me what they had seen, that neither Lieutenant Barber nor Lieutenant Bishop got clear. The aircraft settled in Iron Bottom Bay, Guadalcanal, so named for the planes and ships that sunk there, both Japanese and American.

For lack of a body, it was especially difficult for my aunt and uncle to accept the death of their only child. Unhappily there was never any closure for them. My brother and I are the only male children of our generation to continue that branch of the Bishop family tree.

Mother was a packrat. For anyone schooled by the Great Depression and rationing of World War II it would be against nature to be otherwise. But she was an extreme case, a blue ribbon, champion saver. With her, recycling became an obsession. Every closet, or any piece of furniture with drawers, held its share of precious and trivial items. String, aluminum foil, plastic bags, and carefully washed jars adorned the countertops—waiting to serve her needs one more time. I'm sure that if it were physically possible, she would have found a way to wash and iron Kleenex.

Emma Reeve Bishop
"Little Miss New Yawk"

Mom thought of her possessions with as much reverence as she did the family members. To mention any move where something had been left behind would bring her to tears, but at eighty-eight she had to move into a one-bedroom apartment. I sent her off for the day so it would not be necessary to witness disposition of the excess. When the movers picked up her desk, it groaned, and gave up. On the floor—in brown envelopes, shoeboxes, and packets tied with string—lay all of her treasures. I sighed, packed them into boxes, and moved them to my garage where it could all be sorted. Weeks later I realized I'd become custodian of a vast collection of family memorabilia: pictures, greeting cards, birth or death certificates, diplomas, newspaper clippings, and more. All accumulated by the last survivor of her generation.

Eric George Bishop,
At His Oklahoma City Home

I've had parents most of my life. Dad was my mentor at times and always someone with whom I could easily converse. I not only had a father but a best friend. He suffered a leg injury at age ninety-two years and eleven months, and infection caused his death a week later, mother and dad had been married sixty-three years. She survived him by ten years, and also died within days of her ninety-third birthday. As I had commented to him years earlier, their move to the easy-going laid-back pace of Oklahoma City did extend their lives by years.

Because of his disabling injury while working for the city, dad had received an award of Worker's Compensation and Retirement at age forty-seven. New York City paid him half salary for forty-six years. Because his official retirement age was so early, the city wanted to verify now and then that he was alive and cashing his own checks. Twice they hired an agent here in Oklahoma City, to take affidavits and photographs for proof. It pleasures me to say that I've known someone who not only beat the retirement system, but beat the hell out of it.

The Soviet counterpart to Werner Von Braun remained unidentified for years except by the title of, "The Engineer." The USSR leadership feared that he might be assassinated in an attempt to slow their program. After he died during a routine surgery, it was revealed that he was a Russian, an aerodynamics engineer named Sergei Pavlovitch Korelev, sometimes spelled Gorelev. Like Von Braun, he had been arrested on the charge of being too interested in the non-military use of rockets. Ironically, he died because the specialists who could normally have intervened weren't available; his surgery was being done on a weekend so he wouldn't miss time during the workweek.

Korelev's successor was a Russian general. He and numerous of his staff died two years later when an N-1 multicluster rocket, the equivalent of our Saturn-V, blew up at launch. The general had moved his chair close to the rocket in order to demonstrate his faith in the design; his staff didn't dare to show less belief, so they also moved forward—to their misfortune. Another explosion of an N-1, in summer 1969, was what some authorities claim to be the disaster that ended the Soviet's opportunity to reach the moon before us.

In a recent television debate discussing the quality of education available in high schools today, one educator cited Brooklyn Tech on his short list of the three best high schools in the eastern part of the country. During my years in industry, I've become well aware how far-sighted and progressive the mathematics and science programs were at Tech when I attended that school. It was like getting six years of math and science in four.

Employment at Kessell Silverplaters gave me an appreciation for job safety. That, plus time in apprentice machine shops, showed me that I don't enjoy the physical aspects of work as much as the mental part. Don't get me wrong; I consider work an essential in life, not a four-letter word. It's what puts bread and butter on the table every day, but if I like, or enjoy what I'm doing, it isn't hard work. Life will pay whatever wage we ask of it.

For reference as to the prices I've cited—in 1966, when I started the underwater programs, Beefeater Gin cost $1.95 a fifth in St. Thomas, and the local rum was $0.75. Average income in the U.S. was $6899. A new car cost $2653; an average house was $14,175. A loaf of bread was twenty-two cents, and thirty-two bought a gallon of gasoline. Dow Jones' index didn't pass 1000 until 1972, and taxes weren't the burden they seem today

After the Freedom-7 Mercury capsule sank, a flotation collar was designed under the guidance of a NASA program director named Milt Heflin. He is a fellow alumnus from the University of Central Oklahoma, in Edmond, Oklahoma. When the Edmond Historical Society hosted a NASA space and moon rock exhibition, it was my pleasure to be present though I had no opportunity to converse with Milt. A mutual friend in Edmond, Linda Jones, has since introduced me to him.

NASA never did contract with General Electric, or anyone else, for underwater astronaut training; they continue to do it for themselves.

Huge tanks and special facilities have been built in Huntsville, Alabama, and Houston, Texas. In the year 2000, Houston's underwater training activities were moved from the main space campus to a nearby center specializing in that aspect of training. Visitor tours through the new facility started in 2001. Their underwater trainees continue to use an air hose attached to their suit as the means of pressurizing the suit and receiving breathing air, and to wear those bandoleers of lead weights.

I've never known any other visionary quite like Fred Parker. With his ideas seeded in the clouds, and my feet rooted in the ground of reality, we worked well together. For three of the best years of work that anyone could ask for, we made things happen.

John Du Pont, who helped Fred, Carl and the others to get into their own business was the same man cited in the newspapers when he shot and killed an Olympic wrestler who was living and training on Du Pont's Pennsylvania estate. Life crosses our paths with strange bedfellows.

Despite the GE underwater team's efforts in manned spaceflight safety, namely MOOSE program, no individual emergency device has been provided on any launch to date. NASA is building an X-38 emergency vehicle; in the summer of 2000, Wanda and I saw one assembled at the Houston Space Center, and it reminded me of those 'lifting body' vehicles that I'd studied for Advanced Engineering. Sometime in the future, it will be mounted on the International Space Station, which has a crew of three. No doubt the date was coincidental, but the first drop test of an empty X-38 was the same day that the first occupants of the Space Station began a three-month residence, November 3, 2000.

It took a few years for me to understand and accept why the MOL program was cancelled, but not now. Improvements to cameras, film, optical instruments, robotics, and satellite communications made a human being totally unnecessary to accomplish the same things. Look at the photos taken from flyby-satellites, or the Hubbell telescope, and the Hubbell has been repaired or updated by Space Shuttle astronauts twice in the last few years. Larger lenses are in the planning.

I met Carl Beers again, in Oklahoma City. Though I don't know the circuitous route he took, he was transferred here to work with our Human Resources group before his retirement. Another, who came to OKC from Pittsfield, was Bill Kennedy, an engineering manager. I was near the end of my apprenticeship when Bill began the program; he completed the course and went directly to college. Hilliard, or Hilly, Paige, Vice President of Missile and Space Division, transferred to Phoenix with GE Information Systems; I did see him on several of his Oklahoma City visits.

Because of mergers, acquisitions, and the resultant reorganizations, General Electric no longer owns any of the departments or divisions where I've worked. Pittsfield Power and Distribution Transformer Departments are now part of Westinghouse, whom we sometimes had referred to as 'The Enemy.' That company is in turn part of Siemens Company, who is based in Munich, Germany.

GE Missile and Space Department, which started in the A&P Company warehouse, became a division with three departments. By 1969 thirteen thousand employees worked in twenty-six buildings around the Delaware Valley, but that number was cut in half within a year of the MOL cancellation. During the defense industry merger frenzy following the collapse of the Soviet Union in 1989, it was one of several divisions sold to Martin-Marietta Corporation. That sale included the Missile and Space Division, in Philadelphia and King of Prussia, as well as the Naval Ordinance Department in Pittsfield and Dalton, Massachusetts.

As mentioned earlier, General Electric's computer operations were sold to Honeywell in September 1970, ten months after I arrived in Oklahoma City. Through a high of thirty-six hundred employees and a low of one thousand, I managed to remain. I have a larger number of friends from work than I have people in my family, and miss the daily camaraderie that we shared. There was always an excitement for me in the creation of something new, something of value.

Ironically, under a different political structure in the year 2001, the Justice Department was going to allow GE to acquire Honeywell for forty-one billion dollars; not only computers, but the whole corporation. However, European Union squashed that acquisition on the same grounds as the Justice Department had done to GE in the past; once again, someone felt that the company would have too much clout. This time it would have been a problem for the European aircraft industry, not computers. I'd been in Oklahoma City for months before the one-thousandth disk drive was built; today, I've been told by a source at Seagate that, worldwide, the company produces close to one hundred fifty thousand disk drives each day. Those drives have decreased in size until they now fit on the palm of a hand—some can be held in two fingers.

1999
Wren 5-1/4 inch Hard Drive
2.5 pounds, 40 gigabytes

Disk Drive Comparison—30 years

1969
DSU160C w/Removable Media Pack
250 pounds, 7.5 megabytes

For 40 years areal density increased by 38 percent compound growth rate, and drive capacity doubled every couple of years. In the early '90's, with more sensitive heads, density increased by 60 percent, and capacity doubled every 18 months. By 1997, a density increase of 120 percent allowed doubling of new drive capacity every 9 months. Areal density figures are now in excess of 10 gigabytes, and a recent projection indicates that drives of the future will see 10 terrabytes.

When Honeywell purchased the department in 1970, the word went out to spell our product as "disk," not "disc." Honeywell's public relations firm had been using "disk," and since HISI was the survivor of that sale, everyone tried to comply with the spelling. But old habits die hard, and other companies didn't go along, so it is still seen both ways in the industry. However, I think Honeywell had a valid point—everyone spells diskette with a "k." From a high of more than two hundred companies producing disk drives, there are twenty-three survivors today.

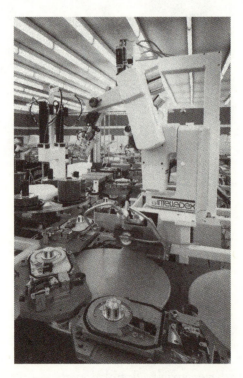

Typical Clean Room with Robotic Assembly
5-1/4 inch Disk Products

For someone raised in the era of science-fiction pulp magazines, and Buck Rogers comics, it's been an exciting, fulfilling career—a dream job. I've worked on satellites, flyby missions or landers for Mars, and the Space Shuttle. Many of the programs I worked on have been implemented, and come to fruition.

Though I worked on GE's proposal for the Voyager, I can't claim any credit for the performance of the final version. As of now the two Voyagers have been in flight for twenty-five years. Voyager 1 is 7.8 billion miles from earth and Voyager 2 is 6.3 billion. They were intended to visit Jupiter and Saturn, but they performed so well that their mission was extended to other heavenly visitations. They passed the last of the nine planets in our solar system ten years ago, and are still headed outward from the sun. Whereas communication to the moon is delayed by 2.3 seconds each way, communications with Voyager 1 is delayed more than twenty-three hours; Voyager 2 takes more than eighteen hours.

The most successful of our underwater testing programs was important for the Skylab, America's first space station. Its actual space launch was in 1973, and successive crews occupied that station. Six years later, on July 11, 1979, segments of that unit burned while reentering earth's atmosphere somewhere east of Australia. Large chunks landed in the Pacific Ocean. A second Skylab vehicle was built but never flown. I've always wondered why we never followed up that first magnificent achievement; the Soviet Union used their Mir station for years. But our remaining unit is in the National Air and Space Museum, and a full size mockup of Skylab space station can be seen at Houston's Space Center.

Internationally, some countries and scientific groups consider the edge of space to be one hundred kilometers, or sixty-two miles high. But in 1953, when Crossfield piloted the X-15 rocket plane the boundary was considered to be fifty miles. By special act of Congress Yeager, Crossfield, and other X-rocket airplane pilots who rode to the fifty mile threshold of space are now recognized as astronauts.

The collapse of the USSR in 1989 has caused Russia to collaborate with the United States on space missions. Economic problems plague the former Soviet Union, and cause delays, but the U.S. has shown a willingness to work with the Soviets. The two countries are cooperating in the creation and operation of the International Space Station, formerly Space Station Alpha. Endeavor space shuttle delivered a giant robot arm to aid in the next level of station assembly. Crewmembers and Mission Commanders from both countries collaborate on missions that John F. Kennedy once conceived but didn't have time to achieve. NASA officials resisted the idea then. On April 12th, 2001 NASA celebrated twenty years of space shuttle flights, one hundred and four to that date.

Much as the prize money offered for the first solo Trans-Atlantic flight inspired pilots, Lindbergh, and others, space has its prize also: X-Prize Foundation has been formed in St. Louis, Missouri. They've offered ten million dollars for the first privately funded spacecraft group to fly two manned missions within fourteen days, to a height of sixty-two miles. Several contenders are testing craft at this time.

A once abandoned airfield in Burns Flat is now headquarters for the Oklahoma Space Development Industry Authority; retired U.S. Air Force General Jay Edwards is the executive director overseeing construction of a commercial spaceport. His group is preparing to launch commercial flights into space in 2002. They anticipate the use of manned space planes from that field by 2007.

JFK inspired the successful moon landings, and 400,000 Americans made it happen. Each helped to achieve a piece of the dream, and should share in accolades for its completion. Humankind has received blessings from the doors that were opened. Mars is the next dream, one that may see fulfillment in my lifetime. An Odyssey spacecraft launched toward the red planet has landed. Among its many tasks is the attempt to find signs of water. If water is to be found, it will most likely be frozen, near either or both poles. In 2003, two rovers will be sent to sites selected from review of the Odyssey data.

Making comments in 1999, on the thirtieth anniversary of the moon landing, our astronauts declared that if funding had been continued at the same level, we would have colonies on Mars today. Just as JFK did for the moon program, someone in authority needs to declare that Martian goal. There are those who dispute the benefit of space spending, in which case I would propose that monies could also be spent on oceanographic studies. Humans live on only one-seventh of the earth's surface. We need to know more about the remainder, and how it affects our lives. Exposure to underwater projects showed me only a brief glimpse of that medium as a science.

It's no longer possible for a non-degreed applicant to use the grandfather clause and test for a Professional Engineering license as I did. That door is closed in every state. I've worked automated systems, or robotics, and consider myself to be a Machine Designer. No one has ever been laid off because of the machines I've designed; on the contrary, they produce something we weren't able to do before, or do it faster and better. With enhanced production, jobs are more secure—not less.

It has also been my privilege to be part of a digital revolution, one that revolutionized and miniaturized the electronic world. Television, communications, electronic games, solar panels, necklace or wrist band computers—all developing at a phenomenal rate. Industry today works in both miniscule and gigantic measurements: microseconds, micro-inches, megabytes, gigabytes and terrabytes. When I worked on Servo Track Writers, tracks were written that were about one hundred and fifty micro inches apart; today that number has been reduced to ten microinches.

Briefly said, a nanosecond is what it takes for an electron to move across nine inches of wire. Microchips are continually being made smaller while each performs more functions than the last design, and faster. Molecular sized microchips have been created in the laboratory. It will take a few years before they are introduced to everyday products, but it will happen.

Our four children are grown now, and flown the nest. Three have families of their own. Steven, the free spirit of the group, is enjoying scenery in the mountains of New Mexico, though he talks of moving to Florida. Eric is a jeweler in Oklahoma City. Altogether there are five nurses in the family: my spouse Wanda, two of the children—Sheryl, in Texas, and Carol Jean in Oklahoma City—plus Sheryl's son Keith and husband Gene. Retired from the Federal Corrections Department, he didn't want to just sit. Now if I have a hangnail I'm covered for advice.

Dad's prediction of change is more than fulfilled. I repeat his words now, "You ain't seen nothin' yet." Our grandchildren are so at ease with the computer and electronic world that I find myself contemplating the wonders their future holds for them. Whether we like it or not, for good or bad, things will change. Heraclitus, one of the Greek philosophers said, "Change is the only constant in our world."

Life may not be fair; it doesn't have to be. No one has ever guaranteed happiness; even the U.S. government only guarantees the right to pursue happiness. My entire life has been one of adaptation, but I didn't give up on my educational dream. I do admit though that—even if you have education and experience—it never hurts to have the pure dumb luck to be in the right spot at the right time. Be careful what you wish for though. You may get it.

0-595-26816-1

www.ingramcontent.com/pod-product-compliance
Lightning Source LLC
Chambersburg PA
CBHW051231050326
40689CB00007B/877